SAFETY

FOR ALL

SAFETY

FOR ALL

EVERY LIFE COUNTS WHEN IT COMES TO SAFETY

Authored By,

Uday Anant Malgaonkar

WINGS PUBLICATION INTERNATIONAL
WE GIVE WINGS TO YOUR DREAM!

Disclaimer

WINGS PUBLICATION INTERNATIONAL
WE GIVE WINGS TO YOUR DREAM!

Registered Office- 907-Sneh Nagar, Sapna Sangeeta Road,
Agrasen Square, Indore – 452001 (M.P.), India

Website: http://www.wingspublication.com

Email: mybook@wingspublication.com

First Published by WINGS PUBLICATION 2024

Title : Safety For All

Price : Rs. 400 I $ 14

ISBN : 978-93-6006-601-7

LIMITS OF LIABILITY/DISCLAIMER OF WARRANTY

Dedication Declaration

"I am dedicating this book to those unfortunate people who died because of not being aware of their safety."

Writer Uday Anant Malgaonkar

Acknowledgment

I want to sincerely thank every one of these people for being a mentor to me and helping me finish this book.

I am grateful to my teachers for their guidance and invaluable insights.

Thank you to our wonderful administrator for providing a learning and creative environment in our school.

My late parents, Mr. and Mrs. Anandibai Anant Malgaonkar, and their unwavering support, tolerance, and understanding have been largely responsible for my achievement.

I am appreciative of my friends' thoughts and viewpoints that enhanced the book.

I am grateful to all of you for helping to shape this book and improve my educational journey.

-Writer Uday Anant Malgaonkar

Foreword 1

Safety is not just a policy, it's a culture—an ever-evolving commitment that begins with awareness and extends into every action we take. In today's fast-paced world, where risks are inherent in every environment—whether at home, in the workplace, or in public spaces—it is crucial to prioritize safety. This book is designed to serve as a comprehensive guide, emphasizing practical strategies, awareness, and behavior that can prevent accidents and save lives.

Throughout my journey in the organisation, I've encountered countless scenarios that could have been avoided with the right knowledge and precautions. This inspired us to create a resource that not only educates but also empowers individuals to become proactive about their safety. From foundational principles to advanced concepts, this book strives to simplify complex safety topics, making them accessible to all readers.

Whether you are a seasoned professional in the safety industry or someone looking to make your environment safer for yourself and your loved ones, this book is for you. My hope is that it serves as a beacon, guiding readers towards building safer habits and a more secure future.

Let us remember that safety is not a destination, but a journey—one we must all walk with vigilance, responsibility, and care.

Regarding the book author our Mr. Uday Anant Malgaonkar had worked with me during my service in TATA motors and we found him very sincere and hard-working personality. He is always hunger for new learning and educate the others also. We found him expert in safety training and promoting his thoughts to implementing safety across various department with a passion for creating safer environments in the company, he is like to learn new things. Also, they are learning themselves from time to time and imparting their skills to others. It is very much aimed at making them learn new things and how they can benefit us and others to the maximum. It is very proud moment for us that he has completed a post-graduation in Safety (ADIS) at the age of 57 with distinction.

In this book, author has stated his rich experience and knowledge on all about human life and it's thought process to save their life by all means. He is the perfect guide for anyone looking to prioritize safety in their lives.

Nitin Tilak,

Retd GM Infra & Env Tata Motors Ltd, Pimpri Pune, Maharashtra

SME for Utility & Maintenance, Environment & I 4.0

Foreword 2

Welcome to this comprehensive guide to safety, where we explore the essential measures and best practices for ensuring a secure environment in various aspects of our lives. This book is dedicated to promoting awareness and knowledge on home safety, office safety, industrial safety, environmental safety, and fire safety.

As we navigate our daily lives, safety is often an afterthought until an incident occurs. However, by prioritizing safety, we can prevent accidents, protect ourselves and others, and create a culture of responsibility.

Within these pages, you will find valuable insights, expert advice, and practical tips on how to identify and mitigate potential hazards in different settings. Whether you are a homeowner, employee, employer, or simply a concerned individual, this book aims to empower you with the knowledge to create a safer world.

Let us take the first step towards a safer tomorrow, today."

Meet Mr. Uday Anant Malgaonkar, a seasoned safety expert in promoting and implementing safety protocols across various industries. With a passion for creating safer environments, Mr. Uday Anant Malgaonkar has dedicated their career to sharing knowledge and expertise with others.

Mr. Uday Anant Malgaonkar holds a graduate in Mechanical Engineering. He has completed his MBA in Operations and has over 34+ years of industrial experience in the Project Execution / Construction/Engineering /Automobile sector. He has completed a post-graduation in Safety (ADIS) also and has worked with numerous organizations, from small businesses to large corporations, to develop and implement effective safety strategies. Their expertise spans across home safety, office safety, industrial safety, environmental safety, and fire safety.

Through this book, Mr. Uday Anant Malgaonkar aims to share their wealth of knowledge and experience with a broader audience, providing practical guidance and insights to help individuals create a safer world. With their unique blend of expertise and enthusiasm, Mr. Uday Anant Malgaonkar is the perfect guide for anyone looking to prioritize safety in their lives.

Sanjay Jagannath Giri

Joint Director, Directorate of Industrial Safety & Health, Pune

Two Words

Dear Reader,

Today, even the smallest mistakes can lead to significant accidents, often with severe consequences. Regardless of the cost, we must all bear these consequences, which can include the loss of life, significant damage to property, and wasted time. Accidents cause not just physical harm but also psychological and social damage. Therefore, it's crucial to prevent accidents, rather than dealing with their aftermath. What precautions should we take, and what actions should we consider in the event of an accident to minimize the loss? Additionally, how do accidents affect the surrounding community?

It's important to recognize that even if an accident was caused by carelessness, we cannot simply ignore it or dismiss our responsibility. We are all part of a society and have obligations that we must fulfil as conscious citizens. This includes being aware of our rights and responsibilities and helping others to do the same. By sharing our knowledge, we can positively impact others, especially those in underprivileged sections of society. This book is inspired by the idea that small actions can have significant benefits for others.

But what exactly is an accident? An accident is an unexpected event, often sudden, that occurs without intention. It may happen due to carelessness, ignorance, or a combination of factors, leading to unfortunate consequences like injury. While some accidents are beyond our control, many result from a lack of awareness or safety precautions. Accidents are a common part of human life, affecting millions globally each year, leading to injuries, disabilities, and property damage. They can occur anywhere—at home, during travel, at work, or in public spaces. However, with the right precautions and mindfulness, many accidents can be prevented.

This book will provide detailed information about various types of accidents, including domestic, road, electrical, environmental, natural, fire, and child-related accidents. You'll learn how these accidents occur and what measures can be taken to prevent them.

By the end of this book, I hope you will have a better understanding of the differences between accidents and incidents. With this knowledge, you will be able to apply preventive measures in your surroundings, thereby protecting the lives of people, animals, and the environment. Enjoy reading, and I wish you a safe and fulfilling life.

Place – Pune,

Date: 04.04.2024

Writer Uday Anant Malgaonkar

Meet The Author

Uday Anant Malgaonkar is a graduate Mechanical Engineer. He has completed an MBA in Operations and has over 34+ years of industrial experience in the Project Execution / Construction/Engineering /Automobile sector. He has completed a post-graduation in Safety (ADIS) also.

Presently he is working as Senior Manager-Maintenance in M/s Tata Motors Limited, Pune, India. He has travelled to 70 cities in India and about 17 countries for project execution work for the company. He is a columnist, book writer, and poet in Marathi, Hindi, and English languages. Passion for Training. He is a motivational speaker for the youth.

Contact No: +91 9552507823

Home Address: D3/401, Hill view, Golden Park, Phase-1, Beturkar Pada, Kalyan West, Pin code-421301, Maharashtra, India.

Email-Id-Udaymalgaonkar1@gmail.com

About The Book

Safety is an important part of life which people sometimes ignore unless it involves money or personal trouble. In short, "safety for all" means that everyone needs safety to live". Safety means being free from loss, and also free from the risk of future loss. There are many different precautions we can take to stay safe. Safety is a 24-hour-a-day, 7-days-a-week job. Safety requires continuous education to bring awareness about the causes of accidents. A major cause of accidents at work is workers engaging in unsafe practices. Statistics from safety officials show that nine out of ten injuries are the result of unsafe acts by workers, which is why it is important to discuss the causes of unsafe acts. Safety is a combination of attitude and behaviour; that is why any person with the right attitude and behaviour can reduce accidents and incidents. Habits can be changed by controlling the way one thinks .... Behaviour-Based Safety.... such concepts and much more can be read in this book...

Index

Part 1

Safety at Home

PART I

1. SAFETY AT HOME

1.1 Newborn Baby Safety

General Safety Overview:

Ensuring the safety of a newborn requires constant vigilance and a secure environment. From proper sleep positions to safe handling, every detail matters when it comes to protecting a baby.

Safe Sleep Practices:

To reduce the risk of Sudden Infant Death Syndrome (SIDS), always place your baby on their back to sleep, in a crib with a firm mattress and no loose bedding or toys.

Handling and Care:

Always support a newborn's head and neck when lifting or carrying them, as their muscles are not yet strong enough to hold up their head.

Bathing Safety:

Never leave a newborn unattended during bath time. Ensure the water temperature is warm but not hot, and use a thermometer to check before placing the baby in the water.

Safe Feeding:

Whether breastfeeding or bottle-feeding, make sure the baby is in an upright position to prevent choking. Always supervise feeding times and burp the baby frequently.

Car Seat Safety:

Ensure your newborn is properly secured in a rear-facing car seat that meets safety standards. The seat should be installed in the back seat of the vehicle, away from airbags.

Creating a Safe Environment:

Keep the baby's sleeping and play areas free of small objects, cords, and other potential choking hazards. Regularly inspect toys and equipment for any signs of wear or damage.

Visitor Guidelines:

Limit visitors in the first few weeks and ensure anyone who holds the baby has clean hands. Newborns have developing immune systems and are more susceptible to infections.

1.2 Child Safety

Here are some sentences and topics related to child safety:

General Child Safety:

Child safety encompasses a wide range of precautions to protect children from harm at home, school, and in public.

Safety at the house:

Childproof your home by securing heavy furniture to the walls, covering electrical outlets with safety caps, and keeping sharp objects, cleaning supplies, and medications locked away and out of reach. Install safety gates at the top and bottom of stairs to prevent falls, and use corner guards on furniture to protect against bumps and bruises. Regularly inspect your home for potential hazards as your child grows and becomes more mobile."

Safety in the water tank:

Never leave children unattended near any body of water, whether it's a bathtub, pool, pond, or even a small bucket, as drowning can happen quickly and silently. Additionally, ensure they understand and follow water safety rules, such as not swimming alone, avoiding deep water, and always wearing life jackets when boating or playing in open water. Constant supervision and clear guidelines are key to preventing water-related accidents.

Road Safety:

Teach them to trust their instincts and if they feel uncomfortable or threatened. They need to understand the difference between safe adults—such as teachers, police officers, or family friends—and those who might pose a danger.

Social media Safety:

Educate them about the risks of sharing personal information, such as their name, address, or school, with strangers online. Encourage open communication about their online experiences, including any interactions that make them feel uncomfortable or unsafe. "

Safety on fire case:

Teach children about fire safety in the home. Create and regularly practice a family escape plan, making sure everyone knows at least two ways to exit each room and a designated meeting spot outside."

Common Safety:

Encourage them to use equipment that is suitable for their age and size, and teach them about taking turns, sharing, and playing safely with others.

Emergency Preparedness:

Teach your child how to dial emergency numbers and what to do in case of different types of emergencies, such as getting lost or encountering someone who is hurt.

Here are some important points and sentences related to choking and suffocation prevention:

Choking Prevention:

General Awareness:

Choking is a leading cause of injury among young children, making it crucial for parents and caregivers to be vigilant about what children put in their mouths.

Safety about feeding the food:

Fruits are always cut into small pieces, especially for toddlers.

Safety of Toys:

Always determine if a toy is small enough to pose a risk.

Supervised Mealtime:

Always supervise children while they're eating, ensuring they sit down and chew food thoroughly before swallowing. Encourage slow eating to minimize the risk of choking.

Educating Children:

Teach children not to run, play, or lie down while eating, as this increases the risk of choking. Encourage them to chew food completely before swallowing.

Prevention of Suffocation:

Safe Sleeping method:

To prevent suffocation, place proper clothes on their backs to sleep in a wooden box with a firm mattress and avoid placing pillows, blankets, stuffed animals, or bumper pads in the wooden box.

Wooden box Safety:

Ensure that the wooden box's slats are no more than 2 3/8 inches apart to prevent the baby's head from becoming trapped. Avoid using older cribs that may not meet current safety standards.

Plastic Bags and Balloons:

Keep plastic bags, deflated balloons, and other potential suffocation hazards out of reach of children. These items can easily cover a child's nose and mouth, restricting their ability to breathe.

Safe Sleep for Toddlers:

Transition toddlers to a bed when they can climb out of their crib to reduce the risk of entrapment and suffocation. Remove any large toys or stuffed animals that could be used to climb out.

Monitor Play Areas:

Regularly inspect your child's play areas for items that could pose a suffocation risk, such as small parts, cords, or loose bedding. Ensure that play tents, tunnels, and other enclosed spaces have adequate ventilation.

Clothing Safety:

Avoid dressing children in clothes with drawstrings or cords that could become entangled and cause suffocation or strangulation. Opt for safer alternatives like snaps, buttons, or Velcro.

Emergency Response:

Learn basic first aid and CPR techniques so that you're prepared to act quickly if a child is choking or suffocating. Early intervention can save lives in critical situations.

1.3 Home Safety of senior citizens.

As people age, they often face an increased risk of accidents in the home due to factors such as decreased mobility, impaired vision, and slower reaction times. For senior citizens, the home, which should be a place of comfort and security, can unfortunately become a place where accidents are more likely to occur.

Common Hazards and Preventative Measures

1. To minimize the risk of falls:

Install grab bars and handrails: Place them in bathrooms, along staircases, and in other areas where support may be needed.

Ensure good lighting: Proper lighting in hallways, staircases, and rooms can help prevent missteps. Consider using nightlights in bathrooms and bedrooms.

2. Fire Hazards:

Fires pose a significant risk to seniors, who may have a slower response time or difficulty evacuating quickly.

Avoid unattended cooking: Seniors should never leave the stove or oven unattended while cooking. Using timers and auto-shutoff devices can help prevent accidents.

Check heating equipment: Regularly inspect heaters and fireplaces for any signs of wear or malfunction, and ensure they are used safely.

3. Medication Safety:

Managing medications can be challenging for seniors, especially those with multiple prescriptions.

Use a medication organizer: A weekly or daily pill organizer can help seniors keep track of their medications and prevent accidental overdoses or missed doses.

Clear labelling: Ensure that all medications are clearly labelled with the correct dosage and instructions. It may be helpful to use large-print labels.

Regular reviews: Periodically review medications with a healthcare provider to avoid harmful interactions and ensure that the prescriptions are still necessary.

4. Safety in bathroom:

Use a shower chair: A sturdy, non-slip shower chair can provide stability while bathing.

Safety precaution

Secure doors and windows: Install deadbolt locks on all doors and ensure that windows have secure locks.

Use a peephole or security camera: This allows seniors to see who is at the door before opening it.

Be cautious with strangers: Seniors should be advised not to open the door to unknown visitors or share personal information over the phone with unsolicited callers.

1. Home Modifications: Consider making home modifications that can enhance safety while promoting ease of movement and accessibility. Examples include installing ramps, widening doorways, and lowering countertops for wheelchair users.
2. Assistive Devices: There are many assistive devices available that

can help seniors perform daily tasks more safely. For example, reachers and grabbers can assist with picking up items, and talking clocks or medication reminders can help those with vision impairments.

3. Emergency Response Systems: Personal emergency response systems (PERS) allow seniors to call for help with the push of a button, providing peace of mind to both the seniors and their families.

Encouraging Awareness and Education

Education is a key component in preventing accidents.

1. Regular Safety Assessments: Periodically review the home for potential safety risks and make necessary adjustments. This could involve rearranging furniture to create clear pathways, checking the stability of handrails, or replacing worn-out equipment.
2. Awareness Programs: Local community centres, healthcare providers, or senior organizations often offer programs or workshops on home safety. Participation in these programs can provide valuable information and resources.
3. Communication: Open communication between seniors, their families, and caregivers is essential in identifying and addressing safety concerns. Regular check-ins can help ensure that seniors are managing well and that any issues are promptly addressed.

By taking proactive steps to address these common safety concerns, we can help ensure that senior citizens can continue to live comfortably, independently, and safely in their own homes.

1.4 Safety at Home.

Home safety is a critical topic, as our homes should be safe havens where we feel secure and comfortable. However, various risks can lead to accidents or other dangers if not properly managed. Look at the sentences below.

1. Poison

Label All Containers at home.

2. Security Measures

Lock Doors and Windows: Always lock doors and windows when leaving the house or going to bed.

3. Safe Sleeping Environments: For infants, ensure that cribs are free of soft bedding, toys, and pillows to reduce the risk of Sudden Infant Death Syndrome (SIDS).

4. Window Guards: Install window guards to prevent children from falling out, especially in multi-story homes.

5. Pet Safety

Secure Hazardous Items: Keep harmful substances, such as certain foods, chemicals, and plants, out of reach of pets.

Safe Spaces: Provide pets with a safe and comfortable space where they can retreat, especially during stressful situations like thunderstorms or fireworks.

6. Water Safety

Supervise Children: Always supervise young children around water, whether it's a bathtub, pool, or natural body of water.

Pool Safety: Install a fence around swimming pools and use pool covers. Ensure that pool gates are self-closing and self-latching.

7. Home Maintenance

Regular Inspections: Conduct regular inspections of your home, including the roof, gutters, and foundation, to catch any potential issues early.

Pest Control: Address any signs of pests promptly to avoid infestations that could lead to damage or health risks.

Promoting a Culture of Safety

Making safety a priority in the home involves creating a culture of awareness and prevention. Encourage all household members to be mindful of potential hazards and to take proactive steps to mitigate them. Regularly review safety procedures and update any equipment or practices as needed to ensure that your home remains a safe and secure environment for everyone.

1.5 Success story

The Tale of a Safe Escape: The Rajan Family's Preparedness Pays Off Background

In a quiet neighbourhood in Kochi, Kerala, lived the Rajan family—Arun, his wife Meera, and their two children, Anaya and Rohit. Like many families, they enjoyed the comfort of their home and often thought about safety, especially after hearing about accidents in the news. Arun, an engineer with a keen sense of responsibility, believed in being prepared for any situation. He regularly discussed safety measures with his family, emphasizing the importance of being ready for emergencies, particularly fires, which can occur unexpectedly.

The Incident

One night in the summer of 2022, as the Rajan family was sleeping, a short circuit in the kitchen sparked a fire. The flames quickly spread to the curtains and began engulfing the kitchen. The family had installed smoke detectors in key areas of the house, and the alarm system went off as soon as the smoke began to fill the kitchen. The loud beeping woke Arun and Meera immediately.

The Response

Thanks to Arun's insistence on practicing emergency drills, the family knew exactly what to do. They had a well-thought-out plan in place, which they had rehearsed several times. Each family member had a designated role in case of an emergency.

1. Arun's Role: Arun was responsible for quickly assessing the situation and guiding the family out of the house. He

immediately checked on the children and ensured that everyone was awake and alert.

2. Meera's Role: Meera was in charge of grabbing the emergency kit, which contained essential items like a flashlight, first aid supplies, and important documents. She also had a list of emergency contacts, including neighbours and local authorities.
3. Children's Role: Anaya and Rohit knew to stay low to avoid inhaling smoke and to follow their parents' instructions without panic. They had been taught to crawl if necessary and to cover their mouths with a cloth to filter out the smoke.

Arun quickly led the family out of the house through the safest exit, which they had predetermined during their safety drills. The fire was already spreading, but because of their preparation, the family was able to evacuate the house in less than two minutes.

The Aftermath

Once outside, the Rajan family immediately called the fire department. The fire brigade arrived promptly and managed to control the fire before it spread to neighbouring homes. Although the kitchen and parts of the living room were damaged, the rest of the house was saved, and most importantly, the entire family was unharmed.

The Rajan family's preparedness and quick thinking not only saved their lives but also minimized the damage to their property. Their story became an example in the community, prompting many neighbours to take home safety more seriously. Arun was invited to share his experience at local safety seminars, where he emphasized the importance of regular fire drills, the use of smoke detectors, and having an emergency kit ready.

The Lesson

The Rajan family's story is a powerful reminder that home safety is not just about having the right equipment; it's also about having a plan and being prepared. Their proactive approach to safety—installing smoke detectors, conducting regular drills, and having a clear evacuation plan—was crucial in their successful escape from a potentially deadly fire.

This true story highlights the importance of preparation and how a few simple precautions can make all the difference in an emergency. The Rajan family's experience has inspired many others in their community to adopt similar safety measures, ensuring that more homes are prepared for unexpected events.

1.6 Brainstorm

Here's a 30-question multiple-choice test on home safety:

Home Safety Multiple-Choice Test

1. What should you do if you smell gas in your home?

    a) Light a match to find the leak

    b) Open windows and doors, and leave the house

    c) Ignore it and go back to bed

    d) Use electrical switches to turn off the lights

2. Which of the following is a good safety practice when using electrical appliances?

    a) Using frayed or damaged cords

    b) Unplugging appliances when not in use

    c) Overloading power strips

    d) Using water to clean plugged-in appliances

3. How often should smoke detector batteries be replaced?

    a) Once a year

    b) Every 5 years

    c) Once a month

    d) Twice a year

4. What is the recommended temperature for hot water to avoid burns?

    a) 160°F

    b) 150°F

c) 140°F

d) 120°F

5. Which of these is NOT a safe practice for using a space heater?

   a) Placing it at least three feet away from anything flammable

   b) Leaving it on while you sleep

   c) Turning it off when you leave the room

   d) Plugging it directly into a wall outlet

6. What should you do in case of a grease fire in the kitchen?

   a) Pour water on the fire

   b) Use a fire extinguisher

   c) Cover the pan with a lid to smother the flames

   d) Move the pan to the sink

7. Where is the best place to store cleaning chemicals?

   a) In a child-accessible cabinet

   b) In a locked cabinet out of reach of children

   c) Under the sink with the doors open

   d) On the kitchen counter

8. How should you dispose of old medication?

   a) Flush it down the toilet

   b) Throw it in the trash

   c) Take it to a pharmacy for proper disposal

   d) Bury it in the yard

9. What type of fire extinguisher should be used in a kitchen?

a) Class A

b) Class B

c) Class C

d) Class D

10. How often should you test your smoke detectors?

a) Every six months

b) Once a year

c) Once a month

d) Every two years

11. What is the safest way to exit a home during a fire?

a) Run through the smoke

b) Crawl low under the smoke

c) Jump out of a window

d) Stay inside and wait for help

12. Which of these is a safe way to store a firearm in the home?

a) Loaded and in a bedside drawer

b) Unloaded and locked in a safe

c) Under a pillow

d) On a high shelf

13. When should you clean the lint filter in your dryer?

a) Once a month

b) After every load

c) Once a week

d) When it gets full

14. What is the purpose of a ground fault circuit interrupter (GFCI) outlet?

    a) To save energy

    b) To protect against electrical shock

    c) To improve signal strength

    d) To prevent circuit overloads

15. What should you do if your carbon monoxide detector goes off?

    a) Open all windows and stay inside

    b) Ignore it, it's probably a false alarm

    c) Leave the house immediately and call emergency services

    d) Turn off the detector and go back to sleep

16. Which of the following is NOT a safety tip for preventing falls in the home?

    a) Keep floors clutter-free

    b) Use rugs without non-slip backing

    c) Install handrails on stairs

    d) Ensure good lighting throughout the home

17. Where should you install carbon monoxide detectors in your home?

    a) Only in the kitchen

    b) In the basement

c) Outside sleeping areas and on every level of the home

d) In the bathroom

18. What is the safest way to use a ladder?

a) Lean it against an unstable surface

b) Stand on the top step

c) Place it on a level, solid surface

d) Use it in the rain

19. What should you do if you experience an electrical shock from an appliance?

a) Keep using the appliance

b) Unplug the appliance and have it checked by a professional

c) Ignore it, it's probably nothing

d) Use the appliance with wet hands

20. What should you do with a frayed electrical cord?

a) Continue using it carefully

b) Tape it up with electrical tape

c) Replace it immediately

d) Ignore it

21. Which of the following is the most common cause of home fires?

a) Faulty wiring

b) Unattended cooking

c) Smoking in bed

d) Candles left burning

22. How far away should candles be kept from anything flammable?

a) 6 inches

b) 1 foot

c) 2 feet

d) 3 feet

23. What is the recommended number of smoke alarms in a home?

a) One on every level of the home

b) One in the kitchen

c) One in the living room

d) One in each bedroom and on every level of the home

24. Which of these practices can help prevent poisoning at home?

a) Storing chemicals in food containers

b) Labelling all chemicals clearly

c) Leaving pills on the counter

d) Mixing cleaning products

25. How should you handle a wet electrical appliance?

a) Plug it in to dry it out

b) Turn it on to check if it still works

c) Let it dry completely before use

d) Dispose of it immediately

26. How often should you inspect your home's electrical system?

a) Every 20 years

b) Every 10 years

c) Every 5 years

d) Every 2 years

27. What is the safest way to use a knife in the kitchen?

a) Use a dull knife to prevent cuts

b) Cut towards yourself for better control

c) Keep your fingers close to the blade

d) Use a sharp knife and cut away from your body

28. How should you treat a minor burn?

a) Apply ice directly to the burn

b) Run cool water over the burn

c) Cover it with butter

d) Pop any blisters

29. What should you do if you see a child playing with matches?

a) Ignore it

b) Teach the child about fire safety and take the matches away

c) Let the child play under supervision

d) Allow the child to keep the matches

30. Which of these is a safe practice when storing heavy items?

a) Store them on high shelves

b) Store them at eye level

c) Store them on lower shelves to prevent them from falling

d) Store them in the attic

Answers:

1. b) Open windows and doors, and leave the house.
2. b) Unplugging appliances when not in use
3. d) Twice a year
4. d) 120°F
5. b) Leaving it on while you sleep
6. c) Cover the pan with a lid to smother the flames
7. b) In a locked cabinet out of reach of children
8. c) Take it to a pharmacy for proper disposal
9. b) Class B
10. c) Once a month
11. b) Crawl low under the smoke
12. b) Unloaded and locked in a safe
13. b) After every load
14. b) To protect against electrical shock
15. c) Leave the house immediately and call emergency services
16. b) Use rugs without non-slip backing
17. c) Outside sleeping areas and on every level of the home
18. c) Place it on a level, solid surface
19. b) Unplug the appliance and have it checked by a professional
20. c) Replace it immediately
21. b) Unattended cooking
22. d) 3 feet

23. d) One in each bedroom and on every level of the home
24. b) Labelling all chemicals clearly
25. d) Dispose of it immediately
26. c) Every 5 years
27. d) Use a sharp knife and cut away from your body
28. b) Run cool water over the burn.
29. b) Teach the child about fire safety and take the matches away.
30. c) Store them on lower shelves to prevent them from falling.

Part 2

Safety at Workplace

PART II

2. SAFETY AT THE WORKPLACE

2.1 Introduction to Safety in the Workplace.

Safety in the workplace is a crucial aspect of any organization, as it directly impacts the health, well-being, and productivity of employees. A safe work environment not only protects workers from accidents, injuries, and illnesses but also fosters a culture of care and responsibility, enhancing overall morale and efficiency. Workplace safety encompasses a broad range of practices, including the use of protective equipment, adherence to safety protocols, proper training, and the identification and mitigation of potential hazards.

In today's dynamic work environments, safety is not just the responsibility of management but a shared commitment among all employees. Effective safety programs involve continuous education, open communication, and the active participation of everyone in the workplace. By prioritizing safety, organizations can reduce the risk of workplace incidents, comply with regulatory standards, and ultimately create a healthier, more productive work environment for all.

2.2 Industrial safety

Industrial safety refers to the management of all operations and events within an industry to protect its employees, equipment, and the environment from harm. It involves the implementation of safety protocols, procedures, and practices designed to minimize risks, prevent accidents, and ensure a safe working environment in industrial settings such as manufacturing plants, factories, construction sites, and other production-oriented environments.

Key Aspects of Industrial Safety:

Hazard Identification: Recognizing potential dangers in the workplace, such as machinery malfunctions, chemical spills, electrical hazards, or structural failures.

Risk Assessment: Evaluating the likelihood and potential impact of these hazards to prioritize safety measures.

Preventive Measures: Implementing controls to mitigate risks, such as installing safety guards on machinery, using proper ventilation systems, and ensuring adequate protective gear for workers.

Training and Education: Providing employees with the necessary knowledge and skills to perform their tasks safely, including proper use of equipment, emergency procedures, and safety protocols.

Compliance with Regulations: Adhering to industry-specific safety standards and legal requirements set by governing bodies, such as the Occupational Safety and Health Administration (OSHA) in the United States.

Emergency Preparedness: Developing and practising emergency response plans for situations like fires, chemical spills, or natural disasters to ensure a quick and effective response.

Continuous Improvement: Regularly reviewing and updating safety practices and procedures to address new risks, incorporate technological advancements, and improve overall safety performance.

Importance of Industrial Safety:

Industrial safety is vital because it helps to:

Protect Workers: By minimizing the risk of injuries, illnesses, and fatalities.

Enhance Productivity: Safe working conditions lead to fewer accidents, reducing downtime and improving overall efficiency.

Reduce Costs: Preventing accidents can significantly reduce costs related to medical expenses, legal liabilities, and compensation claims.

Promote a Positive Work Environment: A strong safety culture fosters trust, morale, and a sense of security among employees.

Comply with Legal Requirements: Ensuring that the company adheres to safety regulations helps avoid penalties and legal issues.

In summary, industrial safety is a comprehensive approach to creating and maintaining a safe work environment in industrial settings, crucial for protecting workers, enhancing productivity, and ensuring regulatory compliance.

2.3 True story about industrial safety

The Tale of a Safe Escape: The Rajan Family's Preparedness Pays Off

Background

In a quiet neighbourhood in Kochi, Kerala, lived the Rajan family—Arun, his wife Meera, and their two children, Anaya and Rohit. Like many families, they enjoyed the comfort of their homes and often thought about safety, especially after hearing about accidents in the news. Arun, an engineer with a keen sense of responsibility, believed in being prepared for any situation. He regularly discussed safety measures with his family, emphasizing the importance of being ready for emergencies, particularly fires, which can occur unexpectedly.

The Incident

One night in the summer of 2022, as the Rajan family was sleeping, a short circuit in the kitchen sparked a fire. The flames quickly spread to the curtains and began engulfing the kitchen. The family had installed smoke detectors in key areas of the house, and the alarm system went off as soon as the smoke began to fill the kitchen. The loud beeping woke Arun and Meera immediately.

The Response

Thanks to Arun's insistence on practising emergency drills, the family knew exactly what to do. They had a well-thought-out plan in place, which they had rehearsed several times. Each family member had a designated role in case of an emergency.

1. Arun's Role: Arun was responsible for quickly assessing the situation and guiding the family out of the house. He immediately checked on the children and ensured that everyone was awake and alert.
2. Meera's Role: Meera was in charge of grabbing the emergency kit, which contained essential items like a flashlight, first aid supplies, and important documents. She also had a list of emergency contacts, including neighbours and local authorities.
3. Children's Role: Anaya and Rohit knew how to stay low to avoid inhaling smoke and follow their parents' instructions without panic. They had been taught to crawl if necessary and to cover their mouths with a cloth to filter out the smoke.

Arun quickly led the family out of the house through the safest exit, which they had predetermined during their safety drills. The fire was already spreading, but because of their preparation, the family was able to evacuate the house in less than two minutes.

The Aftermath

Once outside, the Rajan family immediately called the fire department. The fire brigade arrived promptly and managed to control the fire before it spread to neighbouring homes. Although the kitchen and parts of the living room were damaged, the rest of the house was saved, and most importantly, the entire family was unharmed.

The Rajan family's preparedness and quick thinking not only saved their lives but also minimized the damage to their property. Their story became an example in the community, prompting many neighbours to take home safety more seriously. Arun was invited to share his experience at local safety seminars, where he emphasized the importance of regular fire drills, the use of smoke detectors, and having an emergency kit ready.

The Lesson

The Rajan family's story is a powerful reminder that home safety is not just about having the right equipment; it's also about having a plan and being prepared. Their proactive approach to safety—installing smoke detectors, conducting regular drills, and having a clear evacuation plan—was crucial in their successful escape from a potentially deadly fire.

This true story highlights the importance of preparation and how a few simple precautions can make all the difference in an emergency. The Rajan family's experience has inspired many others in their community to adopt similar safety measures, ensuring that more homes are prepared for unexpected events.

2.4 Emergency preparedness

Emergency preparedness is crucial in any industry because it ensures that an organization is ready to effectively respond to unexpected incidents or disasters, minimizing the impact on human life, property, and the environment.

Here's why emergency preparedness is vital:

1. Protection of Human Life:

Immediate Response: A well-prepared emergency plan enables quick and effective action during emergencies, significantly reducing the risk of injury, illness, or loss of life among employees and others on-site.

Minimized Panic: Preparedness helps employees remain calm and act rationally in a crisis, as they know the procedures to follow, reducing the chances of panic and confusion that can lead to further harm.

2. Minimization of Property Damage:

Damage Control: Effective emergency preparedness measures can limit the extent of damage to facilities, equipment, and inventory. For example, fire suppression systems, proper storage of hazardous materials, and timely evacuation can prevent or minimize damage.

Business Continuity: By protecting critical assets, emergency preparedness supports quicker recovery and business continuity, reducing downtime and financial losses.

3. Environmental Protection:

Preventing Environmental Disasters: Industries, particularly those dealing with hazardous materials, must be prepared to handle spills, leaks, or explosions to prevent contamination of the environment. Emergency preparedness helps contain such incidents before they cause widespread environmental damage.

Regulatory Compliance: Being prepared ensures compliance with environmental regulations, which often mandate specific emergency response measures to protect natural resources.

4. Regulatory Compliance and Legal Protection:

Legal Requirements: Many industries are required by law to have emergency preparedness plans in place, including regular drills and training. Non-compliance can result in legal penalties, fines, and shutdowns.

Liability Reduction: A well-documented and executed emergency preparedness plan can protect an organization from legal liabilities by demonstrating due diligence in safeguarding employees and the public.

5. Preservation of Reputation:

Public Confidence: How a company responds to an emergency can significantly affect its reputation. Being prepared and responding effectively can maintain or even enhance public confidence, while a poorly managed response can lead to long-term damage to the company's image.

Stakeholder Trust: Employees, investors, customers, and the

community expect industries to prioritize safety. Effective emergency preparedness reinforces trust and confidence among these stakeholders.

6. Continuity of Operations:

Disruption Minimization: An emergency can disrupt business Operations, leading to significant financial losses. Preparedness ensures that the organization can continue critical operations or quickly resume them after an incident.

Disaster Recovery: Having a disaster recovery plan as part of emergency preparedness enables the organization to restore Operations, recover data, and repair damage efficiently, reducing the long-term impact on the business.

7. Improved Employee Morale and Safety Culture:

Safety Awareness: Regular emergency preparedness training and drills heighten employees' awareness of safety issues and procedures, fostering a stronger safety culture within the organization.

Confidence and Morale: Knowing that their employer is prepared for emergencies gives employees a sense of security, which can boost morale and productivity.

8. Economic and Financial Protection:

Cost Savings: While emergency preparedness requires an initial investment in planning, training, and equipment, it can save money in the long run by reducing the costs associated with accidents, legal actions, and insurance premiums.

Insurance Benefits: Many insurance companies offer lower premiums to organizations with comprehensive emergency preparedness plans, as these reduce the insurer's risk.

9. Responsibility to the Community:

Community Safety: Industries, especially those located near residential areas, have a responsibility to protect not just their employees but also the surrounding community. Effective emergency preparedness plans include communication and coordination with local authorities and emergency services.

Public Health: Proper preparedness helps prevent or mitigate public health crises that could arise from industrial accidents, such as chemical spills, fires, or explosions.

10. Long-Term Sustainability:

Sustainability Goals: Industries that prioritize emergency preparedness are better positioned to achieve long-term sustainability by ensuring that they can operate safely and responsibly, even in the face of potential disasters.

Resilience Building: Preparedness is a key component of building resilience, enabling organizations to withstand and recover from emergencies, thereby contributing to their long-term viability.

2.5 The Hazard Pyramid

The Hazard Pyramid, often referred to as the Heinrich Pyramid or Safety Pyramid, is a concept in safety management that illustrates the relationship between different levels of workplace incidents and their frequency. Developed by H.W. Heinrich in the 1930s, it suggests that for every serious accident, there are many more minor incidents and near misses, and an even larger number of unsafe acts or conditions that precede them. The pyramid is used to emphasize the importance of addressing minor incidents and unsafe behaviours to prevent more serious accidents.

Structure of the Hazard Pyramid:

Base of the Pyramid: Unsafe Acts or Conditions

Definition: These are the underlying unsafe behaviours, practices, or environmental conditions that have the potential to cause harm but have not yet resulted in an incident.

Examples: Not wearing personal protective equipment (PPE), poor housekeeping, using equipment incorrectly, or ignoring safety protocols.

Frequency: These occur frequently and form the base of the pyramid. Heinrich suggested that there are a large number of unsafe acts or conditions for every minor incident that occurs.

Middle of the Pyramid: Near Misses

Definition: Near misses are events where an accident almost happened but was avoided, either by chance or timely intervention.

Examples: A worker slipping on a wet floor but catching themselves before falling, or a machine malfunctioning but not causing injury.

Frequency: Near misses occur less frequently than unsafe acts but are more common than actual accidents. Heinrich's model suggests that for every serious injury, there are numerous near misses.

Upper Middle of the Pyramid: Minor Incidents

Definition: Minor incidents involve accidents that result in minor injuries or property damage but are not severe.

Examples: Small cuts, bruises, or minor equipment damage that does not lead to significant harm or downtime.

Frequency: These incidents are less frequent than near misses but still occur regularly.

Top of the Pyramid: Major Accidents

Definition: Major accidents are severe incidents that result in significant injury, fatality, major property damage, or environmental harm.

Examples: Fatal accidents, serious injuries requiring hospitalization, major fires, or explosions.

Frequency: These are the least frequent occurrences but have the most severe consequences.

Key Concepts of the Hazard Pyramid:

Proportional Relationship: Heinrich's research indicated that for every major accident, there are approximately 29 minor incidents, 300 near misses, and many more unsafe acts or conditions at the base. This proportion is often represented as 1:29:300, though the exact ratios may vary across different industries.

Prevention Focus: The pyramid highlights the importance of addressing unsafe acts and conditions at the base. By identifying and correcting these issues, organizations can prevent near misses and minor incidents, ultimately reducing the likelihood of major accidents.

Proactive Safety Management: The model encourages a proactive approach to safety management, where attention is given not only to accidents that have occurred but also to potential hazards and near misses that could lead to more serious incidents if left unaddressed.

Modern Interpretation and Criticism:

Heinrich's Law: The pyramid is sometimes used to explain "Heinrich's Law," which states that reducing the number of minor incidents and unsafe acts will reduce the number of serious accidents.

Criticism and Evolution: Some modern safety professionals criticize the original Hazard Pyramid for oversimplifying the relationship between different types of incidents and for not always holding in every situation or industry. They argue that focusing solely on the lower levels of the pyramid may not address the root causes of serious accidents. As a result, the concept has evolved, with greater emphasis now placed on understanding and mitigating specific risks and hazards rather than relying solely on the frequency of incidents.

2.6 Safety standards

Safety standards in the industry are regulations, guidelines, and practices designed to ensure the health, safety, and well-being of workers, equipment, and the environment in various industrial settings. These standards are established by government agencies, international organizations, and industry groups to prevent accidents, injuries, and environmental harm. Here's an overview of key safety standards in different industries:

1. Occupational Safety and Health Administration (OSHA) Standards

Scope: OSHA is a U.S. government agency responsible for setting and enforcing workplace safety standards across various industries, including construction, manufacturing, healthcare, and more.

Examples of OSHA Standards:

General Industry Standards (29 CFR 1910): Covers a broad range of safety requirements, including machine guarding, hazardous materials handling, and electrical safety.

Construction Standards (29 CFR 1926): Focuses on construction site safety, including fall protection, scaffolding, and excavation safety.

Hazard Communication Standard (HazCom): Requires the labelling of hazardous chemicals and the provision of Safety Data Sheets (SDS) to inform workers about the risks and proper handling procedures.

2. International Organization for Standardization (ISO) Standards

Scope: ISO develops and publishes international standards for various industries, including safety management systems, quality, and environmental management.

Examples of ISO Standards:

ISO 45001: An international standard for Occupational Health and Safety Management Systems (OHSMS), providing a framework for managing and improving workplace safety.

ISO 14001: Focuses on environmental management systems, helping organizations minimize environmental impact while ensuring compliance with environmental regulations.

ISO 12100: Provides guidelines for the safety of machinery, including risk assessment and risk reduction techniques.

3. American National Standards Institute (ANSI) Standards

Scope: ANSI oversees the development of voluntary consensus standards in the U.S., covering a wide range of safety topics.

Examples of ANSI Standards:

ANSI Z10: Occupational Health and Safety Management Systems standard, similar to ISO 45001, emphasizing continuous improvement in safety performance.

ANSI Z87.1: Specifies requirements for eye and face protection devices, including safety goggles and face shields.

ANSI/ISEA 107: Standard for high-visibility safety apparel,

commonly used by workers in construction, road work, and emergency response.

4. National Fire Protection Association (NFPA) Standards

Scope: NFPA develops codes and standards to reduce the risk of fire, electrical hazards, and related dangers in various industries.

Examples of NFPA Standards:

NFPA 70 (National Electrical Code or NEC): Sets the standard for safe electrical design, installation, and inspection to protect people and property from electrical hazards.

NFPA 101 (Life Safety Code): Establishes safety requirements for the construction and operation of buildings to protect occupants from fire, smoke, and other hazards.

NFPA 70E: Provides guidelines for electrical safety in the workplace, including requirements for safe work practices, PPE, and hazard assessments.

5. Mine Safety and Health Administration (MSHA) Standards

Scope: MSHA is a U.S. government agency that sets and enforces safety and health standards for mining operations to protect miners from hazards.

Examples of MSHA Standards:

Part 46: Training and retraining requirements for miners, focusing on safety and health topics specific to the mining industry.

Part 50: Requirements for accident reporting and investigation in mining operations.

Part 57: Safety standards for surface and underground metal and nonmetal mines, including ventilation, ground control, and electrical safety.

6. Chemical Safety Standards (e.g., REACH, GHS)

REACH (Registration, Evaluation, Authorization, and Restriction of Chemicals):

Scope: A European Union regulation aimed at improving the protection of human health and the environment from the risks posed by chemicals. It requires companies to register the chemicals they produce or import and provide safety information.

Globally Harmonized System of Classification and Labelling of Chemicals (GHS):

Scope: An internationally agreed-upon system for classifying and labelling chemicals, ensuring that information about chemical hazards is consistent and understandable worldwide.

7. Food and Drug Administration (FDA) Standards

Scope: The FDA is responsible for protecting public health by ensuring the safety, efficacy, and security of drugs, food, and medical devices in the United States.

Examples of FDA Standards:

Current Good Manufacturing Practices (cGMP): Regulations that provide guidelines for the proper design, monitoring, and control of manufacturing processes and facilities in the pharmaceutical and food industries.

Food Safety Modernization Act (FSMA): Focuses on preventing

food safety issues through proactive measures in food production, processing, and distribution.

8. Environmental Protection Agency (EPA) Standards

Scope: The EPA sets and enforces environmental standards in the United States to protect human health and the environment.

Examples of EPA Standards:

Clean Air Act: Regulates air emissions from stationary and mobile sources to protect air quality.

Clean Water Act: Establishes standards for water quality and regulates the discharge of pollutants into U.S. waters.

Resource Conservation and Recovery Act (RCRA): Governs the management and disposal of hazardous and non-hazardous solid waste.

9. Industry-Specific Safety Standards

Aerospace: The Aerospace Standard AS9100 defines quality and safety management requirements for aerospace manufacturers.

Oil and Gas: API (American Petroleum Institute) standards provide guidelines for safe operations, equipment, and practices in the oil and gas industry, including offshore drilling and pipeline safety.

Healthcare: The Joint Commission sets standards for patient safety and quality of care in healthcare organizations, focusing on areas like infection control, medication safety, and emergency preparedness.

10. International Labour Organization (ILO) Standards

Scope: The ILO develops international labour standards, including those related to occupational safety and health, aimed at promoting decent work conditions worldwide.

2.7 Life-saving methods.

Life-saving rules in the industry are specific, critical safety guidelines designed to prevent severe injuries and fatalities. These rules are often based on common causes of accidents in industrial settings and are intended to be non-negotiable, meaning that they must be strictly followed at all times. Here are some common life-saving rules that are widely implemented across various industries:

1. Work at Height Safety:

Use of Fall Protection: Always wear appropriate fall protection equipment, such as harnesses, when working at heights above a certain level (e.g., 2 meters or 6 feet).

Secure Tools and Materials: Ensure all tools and materials are secured to prevent them from falling and causing injury to people below.

Proper Scaffolding: Only use properly erected and inspected scaffolding or ladders.

2. Energy Isolation (Lockout/Tagout - LOTO):

Isolate Energy Sources: Before performing maintenance or repairs on machinery, ensure all energy sources (electrical, mechanical, hydraulic, etc.) are isolated and locked out.

Verify Isolation: Always verify that the energy source is effectively isolated before beginning work.

Use Proper Lockout Devices: Use approved lockout devices and tags to indicate that equipment is isolated and should not be operated.

3. Confined Space Entry:

Obtain a Permit: Never enter a confined space without a proper permit and authorization.

Test the Atmosphere: Test the atmosphere in the confined space for hazardous gases, oxygen levels, and other potential dangers before entry.

Use of Safety Equipment: Always wear appropriate personal protective equipment (PPE), including respiratory protection if necessary, and ensure a standby person is present outside the confined space.

4. Safe Driving and Mobile Equipment Operation:

Seat Belts: Always wear a seat belt when operating vehicles or mobile equipment, such as forklifts.

Speed Limits: Adhere to speed limits and other traffic regulations within the worksite.

No Mobile Phone Use: Do not use mobile phones or other distractions while driving or operating equipment.

5. Control of Hazardous Substances:

Use of Proper Containers: Store and handle hazardous substances in approved containers with appropriate labelling.

Wear PPE: Always wear the necessary PPE, such as gloves, goggles, and protective clothing, when handling hazardous substances.

Follow Handling Procedures: Adhere strictly to safety procedures for handling, transporting, and disposing of hazardous materials.

6. Permit to Work (PTW):

Obtain a Permit: Before starting any high-risk work, such as hot work, confined space entry, or electrical work, obtain a permit to work.

Follow Permit Conditions: Ensure all conditions specified in the permit are met and follow the safety measures outlined in the permit.

Sign Off on Completion: Ensure that the permit is properly signed off once the work is completed, confirming that the area is safe.

7. Lifting Operations:

Use Proper Equipment: Only use cranes, hoists, and other lifting equipment that is rated for the load and has been inspected and maintained.

Plan the Lift: Ensure that all lifting operations are planned, with clear communication and roles assigned to all involved personnel.

Stay Clear of Load: Never stand or walk under a suspended load.

8. Electrical Safety:

De-energize Equipment: Always de-energize and lock out electrical equipment before working on it.

Inspect Tools: Use only insulated tools and inspect them before use.

Stay Clear of Overhead Lines: Maintain a safe distance from overhead power lines when operating cranes or other equipment.

9. Hot Work Safety:

Obtain a Permit: Always obtain a hot work permit before performing tasks such as welding, cutting, or grinding.

Fire Watch: Ensure a fire watch is present during and after hot work to monitor for potential fires.

Remove Flammables: Clear the area of flammable materials before starting hot work.

10. Personal Protective Equipment (PPE):

Wear Required PPE: Always wear the required PPE for the task, such as helmets, gloves, safety glasses, and steel-toed boots.

Inspect PPE: Inspect PPE before use to ensure it is in good condition and provides adequate protection.

Use PPE Correctly: Ensure PPE is used correctly, fitting properly, and worn throughout the task.

11. Substance Use Policy:

No Drugs or Alcohol: Do not consume drugs or alcohol before or during work, as they impair judgment and reaction times.

Report Impairment: If you suspect someone is impaired, report it immediately to a supervisor.

12. Emergency Preparedness:

Know Emergency Procedures: Be familiar with the emergency procedures for your workplace, including evacuation routes, emergency contacts, and the location of emergency equipment like fire extinguishers and first aid kits.

Participate in Drills: Take part in regular emergency drills to ensure you are prepared in the event of an actual emergency.

Report Emergencies: Immediately report any emergency, such as fires, spills, or medical incidents, to the appropriate personnel.

2.8 Safety terminologies

Safety terminologies in the industry are essential for understanding and implementing effective safety practices. Here are 10 key safety terminologies commonly used in various industries:

1. Hazard

Definition: A hazard is a condition, object, or activity with the potential to cause harm, injury, illness, or damage to property or the environment.

Example: Wet floors, exposed electrical wires, and toxic chemicals are common hazards in many industries.

2. Risk

Definition: Risk is the likelihood or probability that a hazard will cause harm or injury, coupled with the severity of the potential consequences.

Example: The risk of slipping on a wet floor increases if the area is not marked with a warning sign, and the potential injury could range from a minor bruise to a serious fracture.

3. Accident

Definition: An accident is an unplanned, unexpected event that results in injury, illness, damage, or loss.

Example: A worker falling from a ladder due to improper use, resulting in a broken arm, is an accident.

4. Incident

Definition: An incident is any event that occurs at a workplace that could or does result in harm. It includes both accidents and near misses.

Example: A chemical spill that does not result in injury but could have caused harm is an incident.

5. Near Miss

Definition: A near miss is an event where an accident was narrowly avoided. Although no one was injured and no damage occurred, the potential for harm was present.

Example: A tool falling from a height and landing near a worker, but without causing injury, is considered a near miss.

6. Personal Protective Equipment (PPE)

Definition: PPE refers to clothing or equipment worn by workers to protect them from hazards in the workplace.

Example: Helmets, gloves, safety goggles, earplugs, and steel-toed boots are examples of PPE.

7. Lockout/Tagout (LOTO)

Definition: Lockout/Tagout is a safety procedure used to ensure that dangerous machines are properly shut off and cannot be started up again before the completion of maintenance or repair work.

Example: Before servicing an electrical panel, the energy source is isolated, locked, and tagged to prevent accidental activation.

8. Safety Data Sheet (SDS)

Definition: An SDS is a document that provides detailed information about a hazardous chemical, including its properties, hazards, protective measures, and safety precautions for handling, storing, and transporting it.

Example: An SDS for a cleaning solvent might include information on first aid measures, fire-fighting instructions, and safe disposal methods.

9. Occupational Health and Safety (OHS)

Definition: OHS refers to the field of study and practice focused on the safety, health, and welfare of people at work. It involves the identification, assessment, and control of workplace hazards.

Example: Implementing ergonomic assessments to prevent repetitive strain injuries in office workers is part of OHS.

10. Safety Culture

Definition: Safety culture is the collective attitudes, beliefs, values, and behaviours of an organization's employees regarding workplace safety. A strong safety culture prioritizes safety in all operations and encourages employees to take responsibility for their own and others' safety.

Example: A company where all employees feel empowered to report safety concerns without fear of retaliation has a strong safety culture.

2.9 Brainstorm

Here's a 30-question multiple-choice test on Industrial safety:

Industrial Safety MCQ Test

1. Which of the following is an example of administrative control in industrial safety?

   a) Installing safety barriers

   b) Providing safety training

   c) Using PPE

   d) Implementing ventilation systems

2. What is the primary purpose of a safety data sheet (SDS)?

   a) To list the price of hazardous materials

   b) To provide information on the safe handling of chemicals

   c) To record workplace accidents

   d) To outline emergency exit routes

3. Which of the following is considered a physical hazard?

   a) Noise

   b) Bacteria

   c) Chemical spills

   d) Stress

4. What does the term "HazCom" refer to?

   a) Hazardous Communication Standard

b) Hazardous Combustion

c) Hazardous Company Policies

d) Hazardous Community

5. The hierarchy of controls suggests that the most effective way to manage hazards is through:

   a) PPE

   b) Administrative controls

   c) Engineering controls

   d) Elimination of the hazard

6. Which of the following is a common fire hazard in industrial settings?

   a) Wet floors

   b) Electrical wiring faults

   c) Broken windows

   d) Excessive noise

7. The primary purpose of a Job Safety Analysis (JSA) is to:

   a) Increase productivity

   b) Identify and mitigate hazards associated with specific tasks

   c) Monitor employee attendance

   d) Evaluate employee performance

8. Which of the following is NOT a component of a typical emergency action plan?

a) Evacuation procedures

b) Alarm systems

c) Budget planning

d) Designated assembly areas

9. A "confined space" in an industrial context is defined as:

   a) A small area with limited access

   b) A space not designed for continuous occupancy

   c) A well-ventilated area

   d) A location with a lot of natural light

10. What is the primary function of a machine guard?

    a) To enhance machine efficiency

    b) To protect the machine from damage

    c) To prevent worker contact with moving parts

    d) To reduce noise levels

11. Which of the following is an example of a biological hazard?

    a) Loud machinery

    b) Hazardous chemicals

    c) Infectious bacteria

    d) Slippery surfaces

12. What is a key component of an effective safety culture?

    a) Punishing employees for accidents

b) Promoting worker involvement in safety decisions

c) Reducing safety meetings to save time

d) Ignoring near-miss incidents

13. Personal fall arrest systems should be inspected:

    a) Annually

    b) Monthly

    c) Before each use

    d) Only after a fall occurs

14. Which of the following is a typical sign of heat stress?

    a) Cold sweats

    b) Dizziness

    c) Increased appetite

    d) Shivering

15. In the context of industrial safety, "LOTO" stands for:

    a) Light Out, Timer On

    b) Lock Out, Tag Out

    c) Level of Task Operation

    d) Local Occupational Training

16. When lifting heavy objects, workers should:

    a) Bend at the waist and lift with their back

    b) Twist their body to lift the object

c) Keep the object close to their body and lift it with their legs

d) Lift quickly to reduce strain

17. Which of the following is NOT a recommended practice when working with electrical equipment?

    a) Wearing insulated gloves

    b) Keeping water away from electrical sources

    c) Grounding electrical tools

    d) Using metal ladders

18. Hearing protection is necessary when noise levels exceed:

    a) 70 dB

    b) 85 dB

    c) 90 dB

    d) 100 dB

19. Which colour is typically used for caution signs in an industrial environment?

    a) Red

    b) Blue

    c) Yellow

    d) Green

20. What is the most common cause of workplace fatalities in the construction industry?

    a) Electrocution

b) Falls

c) Being struck by objects

d) Trenching accidents

21. The "three-point contact" rule is used to prevent:

    a) Slips on wet surfaces

    b) Electrical shocks

    c) Falls from ladders

    d) Machine entanglement

22. Which of the following is NOT a type of machine safeguard?

    a) Fixed guard

    b) Adjustable guard

    c) Automatic stop

    d) Warning sticker

23. In case of a chemical spill, the first action should be to:

    a) Clean up the spill immediately

    b) Evacuate the area

    c) Inform your supervisor

    d) Call the fire department

24. Respirators must be used when:

    a) Dust levels are low

    b) The air is clean

c) There is a risk of inhaling hazardous substances

d) Working in a noisy environment

25. The term "MSDS" has been replaced by:

a) SOS (Safety Operations Sheet)

b) SDS (Safety Data Sheet)

c) HSS (Hazardous Safety Sheet)

d) EDS (Emergency Data Sheet)

26. Safety signs indicating "Prohibition" typically use which colour?

a) Green

b) Blue

c) Red

d) Yellow

27. Which of the following is NOT considered a form of personal protective equipment (PPE)?

a) Earplugs

b) Safety goggles

c) Steel-toed boots

d) Ladder

28. What is the purpose of a safety audit?

a) To reduce employee salaries

b) To identify potential safety improvements

c) To monitor productivity

d) To enhance customer satisfaction

29. Which of the following is a sign of a potential ergonomic issue?

    a) Employees reporting frequent muscle strain

    b) High productivity levels

    c) Low absenteeism

    d) Good posture

30. In an emergency evacuation, employees should:

    a) Use elevators to leave quickly

    b) Follow their supervisor at all times

    c) Evacuate in an orderly manner using the nearest exit

    d) Wait for instructions before leaving

Answers:

1. b) Providing safety training
2. b) To provide information on the safe handling of chemicals
3. a) Noise
4. a) Hazardous Communication Standard
5. d) Elimination of the hazard
6. b) Electrical wiring faults
7. b) Identify and mitigate hazards associated with specific tasks

8. c) Budget planning
9. b) A space not designed for continuous occupancy
10. c) To prevent worker contact with moving parts
11. c) Infectious bacteria
12. b) Promoting worker involvement in safety decisions
13. c) Before each use
14. b) Dizziness
15. b) Lock Out, Tag Out
16. c) Keep the object close to their body and lift it with their legs
17. d) Using metal ladders
18. b) 85 dB
19. c) Yellow
20. b) Falls
21. c) Falls from ladders
22. d) Warning sticker
23. b) Evacuate the area
24. c) There is a risk of inhaling hazardous substances
25. b) SDS (Safety Data Sheet)
26. c) Red
27. d) Ladder
28. b) To identify potential safety improvements

29. a) Employees reporting frequent muscle strain

30. c) Evacuate in an orderly manner using the nearest exit

Part 3

Safety on Road

PART III

3. SAFETY ON THE ROAD

3.1 Brief on Road Safety

Road Safety refers to the measures and strategies designed to prevent road users from being killed or seriously injured in traffic collisions. It encompasses various aspects, including the behaviour of drivers, cyclists, and pedestrians, the design and maintenance of roads, and the safety features of vehicles.

Key Elements of Road Safety:

Driver Behaviour:

Speed Management: Adhering to speed limits is crucial as higher speeds increase the severity of accidents. Enforcement of speed limits through speed cameras and penalties helps reduce speeding.

Avoiding Distractions: Distracted driving, such as using mobile phones while driving, significantly increases the risk of accidents. Awareness campaigns and legislation banning phone use while driving are common measures.

Alcohol and Drug Impairment: Driving under the influence of alcohol or drugs impairs judgment and reaction times. Strict laws, Breathalyzer tests, and sobriety checkpoints are essential for deterring impaired driving.

Pedestrian and Cyclist Safety:

Crosswalks and Sidewalks: Providing safe and well-marked crosswalks, pedestrian signals, and sidewalks helps protect pedestrians.

Bicycle Lanes: Dedicated bike lanes and paths ensure that cyclists have safe spaces to ride, separate from motor vehicle traffic.

Vehicle Safety:

Safety Features: Modern vehicles are equipped with safety features such as airbags, anti-lock braking systems (ABS), electronic stability control, and crash avoidance technologies that help protect occupants and reduce the likelihood of collisions.

Regular Maintenance: Ensuring that vehicles are regularly serviced and well-maintained (e.g., brakes, tyres, lights) is critical to preventing accidents due to mechanical failures.

Road Design and Infrastructure:

Road Layout: Designing roads with safety in mind includes features like proper signage, barriers, guardrails, and well-marked lanes.

Traffic Calming: Implementing traffic calming measures such as speed bumps, roundabouts, and narrower lanes can reduce speeds and enhance safety in residential or high-traffic areas.

Lighting: Adequate street lighting improves visibility for drivers, pedestrians, and cyclists, especially at night.

Law Enforcement and Legislation:

Traffic Laws: Enforcing traffic laws related to speed limits, seat belt use, and road signs is crucial for maintaining order and safety on the roads.

Penalties: Implementing penalties for traffic violations, such as fines, license points, and suspensions, helps deter dangerous behaviours.

Public Awareness and Education:

Safety Campaigns: Public awareness campaigns educate road users about safe driving practices, the dangers of impaired and distracted driving, and the importance of using seat belts and helmets.

Driver Education: Mandatory driver education and training programs help new drivers develop safe driving skills and knowledge of traffic laws.

Importance of Road Safety:

Reducing Fatalities and Injuries: Road safety measures are essential for reducing the number of deaths and serious injuries resulting from traffic accidents.

Economic Impact: Traffic accidents impose significant economic costs due to healthcare expenses, lost productivity, and property damage. Improving road safety can help reduce these costs.

Enhancing Quality of Life: Safe roads contribute to a higher quality of life by allowing people to travel without fear of accidents, improving accessibility, and supporting the smooth functioning of communities.

Conclusion:

Road safety is a shared responsibility that involves government agencies, road users, vehicle manufacturers, and the general public. By adhering to traffic laws, using safety features, and fostering a culture of safety, the risks associated with road travel can be significantly reduced, leading to safer roads for everyone.

3.2 List of Road Safety Rules in India

Road safety rules in India are designed to ensure the safety of all road users, including drivers, passengers, pedestrians, and cyclists. These rules are enforced by the Motor Vehicles Act of 1988 and other related regulations. Here is a list of key road safety rules in India:

1. Obey Traffic Signals and Signs:

Always follow traffic lights (red, yellow, green) and road signs, such as stop signs, yield signs, speed limits, and pedestrian crossings.

2. Adhere to Speed Limits:

Drive within the prescribed speed limits for different areas (urban, rural, highways) as indicated by road signs.

3. Wear Seat Belts:

Both drivers and passengers in the front seats must wear seat belts at all times while the vehicle is in motion.

4. No Use of Mobile Phones While Driving:

Using mobile phones for calls, texting, or any other purpose while driving is prohibited unless a hands-free device is used.

5. No Driving Under the Influence:

Driving under the influence of alcohol or drugs is strictly illegal. The permissible blood alcohol content (BAC) limit is 0.03%.

6. Helmet for Two-Wheeler Riders:

Both the rider and pillion passengers on two-wheelers must wear helmets that meet safety standards.

7. Yield to Emergency Vehicles:

Always give way to emergency vehicles like ambulances, fire trucks, and police vehicles by moving to the side of the road.

8. No Overtaking in Dangerous Zones:

Avoid overtaking in areas where it is dangerous, such as curves, intersections, pedestrian crossings, and railway crossings.

9. Use of Indicators:

Always use indicators to signal your intent to turn or change lanes. Ensure that all vehicle lights are functioning properly.

10. Right of Way to Pedestrians:

Give pedestrians the right of way at pedestrian crossings and always slow down near schools, hospitals, and residential areas.

11. Use of Horn:

Use the horn sparingly and only when necessary to alert others of your presence. Do not use the horn in no-honking zones, such as near schools and hospitals.

12. Maintain Safe Distance:

Maintain a safe distance from the vehicle in front of you to avoid collisions in case of sudden braking.

13. No Jumping Red Lights:

Never jump a red light or try to beat the yellow light. Always stop when the light turns red.

14. Proper Lane Discipline:

Stick to your lane and avoid frequent lane changes. Use the correct lane for your speed and direction.

15. Use of Headlights:

Use headlights during low visibility conditions such as fog, rain, or at night. Avoid using high beams when following or approaching other vehicles.

16. Do Not Overload Vehicles:

Ensure that your vehicle is not overloaded with passengers or cargo beyond its capacity.

17. Proper Vehicle Maintenance:

Regularly maintain your vehicle to ensure that brakes, tyres, lights, and other essential components are in good working condition.

18. No Racing or Reckless Driving:

Avoid street racing or driving recklessly, which endangers your life and the lives of others.

19. Follow Rules at Roundabouts:

Yield to vehicles already in the roundabout and follow the indicated lanes and exits.

20. Respect School Zones:

Drive slowly and carefully in school zones, paying attention to children and crossing guards.

21. Avoid Unnecessary Honking:

Unnecessary honking is considered a nuisance and should be avoided, especially in silent zones.

22. Follow Parking Rules:

Park only in designated areas and ensure that your vehicle does not obstruct traffic or pedestrians.

23. Carry Valid Documents:

Always carry valid documents such as your driving license, vehicle registration, insurance, and pollution under control (PUC) certificate.

24. No Triple Riding on Two-Wheelers:

Do not carry more than one million passengers on a two-wheeler.

25. Respect Road Conditions:

Adjust your driving according to road conditions, such as potholes, wet roads, or roadworks.

Conclusion:

These road safety rules are designed to minimize accidents and ensure the safety of all road users. Strict adherence to these rules, combined with a responsible attitude towards driving, can significantly reduce the risk of road traffic accidents and enhance overall safety on the roads in India.

3.3 Road safety rules for kids

Road safety rules for kids are essential to ensure their safety as pedestrians, cyclists, or passengers. Teaching children these rules from a young age helps them develop good habits and be aware of the dangers on the road. Here are some important road safety rules for kids:

1. Stop, Look, and Listen:

Stop: Before crossing the road, always stop at the curb or edge of the road.

Look: Look both ways—left, right, and left again—to check for oncoming traffic.

Listen: Listen for any sounds of approaching vehicles.

2. Cross at Designated Crossings:

Always use pedestrian crossings, zebra crossings, or footbridges when crossing the road. Never cross between parked cars or at a blind spot.

3. Wait for the Signal:

Only cross the road when the pedestrian light is green or when an adult signals that it is safe to cross.

4. Hold Hands with an Adult:

Younger children should always hold hands with an adult when crossing the road or walking near traffic.

5. Walk, Don't Run:

Always walk, never run, across the road. Running can cause accidents as drivers may not see the child in time.

6. Use Sidewalks:

Walk on the sidewalk or pavement, away from the edge of the road. If there is no sidewalk, walk on the side of the road facing traffic so you can see oncoming vehicles.

7. Be Visible:

Wear bright-coloured or reflective clothing, especially in low light conditions, to be more visible to drivers.

8. No Playing on the Road:

Never play on the road or in the driveway. Always play in safe areas like parks or playgrounds.

9. Be Careful with Driveways:

Watch out for vehicles that may be reversing out of driveways. Make sure the driver sees you before you walk behind a car.

10. Follow the School Bus Rules:

Wait for the school bus to come to a complete stop before approaching it.

Always board and exit the bus in an orderly manner, using the handrail.

After getting off the bus, wait for the bus to move away before crossing the road, ensuring there's no oncoming traffic.

11. Learn Traffic Signs and Signals:

Teach children the meaning of basic traffic signs and signals, such as stop signs, pedestrian crossings, and traffic lights.

12. Bicycle Safety:

Always wear a helmet when riding a bicycle.

Use the correct hand signals to indicate turns and stops.

Ride on the correct side of the road (usually the left side) and follow traffic rules.

Use lights and reflectors on the bicycle when riding at dusk or in the dark.

13. Never Approach a Moving Vehicle:

Don't try to get close to or touch a moving vehicle. Always stay a safe distance away.

14. Don't Stick Hands or Heads Out of Vehicle Windows:

When riding in a car, never stick your hands, arms, or head out of the window.

15. Stay Seated and Buckled Up:

Always sit in the back seat of a vehicle and wear a seat belt. Never stand up or move around while the car is moving.

16. Be Aware of Blind Spots:

Understand that drivers might not see you if you are in their blind spot, especially larger vehicles like trucks and buses. Make sure you are visible before crossing in front of them.

Conclusion:

By teaching and reinforcing these road safety rules, children can be better prepared to navigate the roads safely, whether they are walking, cycling, or riding in a vehicle. Adult supervision and modelling good behaviour are also critical in ensuring kids understand and follow these safety guidelines.

3.4 Road safety rules for students

Road safety rules for students are designed to ensure their safety as they travel to and from school and other activities. These rules cover various scenarios, including walking, cycling, and riding in vehicles. Here's a comprehensive list of road safety rules tailored for students:

1. Use Pedestrian Crossings:

Always cross the road at designated pedestrian crossings, such as zebra crossings or pedestrian signals. If there are no crossings, cross at the corner where traffic is more controlled.

2. Follow Traffic Signals:

Wait for the pedestrian signal to turn green before crossing the road. If there is no traffic signal, make sure the road is clear of vehicles before crossing.

3. Walk on Sidewalks:

Use sidewalks or pavements whenever available. If there is no sidewalk, walk facing oncoming traffic so you can see vehicles approaching.

4. Look and Listen:

Before crossing, stop at the edge of the road, look both ways (left, right, and left again), and listen for traffic. Only cross when it is safe.

5. Avoid Distractions:

Stay focused on the road and avoid distractions such as mobile phones or headphones. Pay attention to your surroundings.

6. Wear Visible Clothing:

Wear bright or reflective clothing, especially in low light conditions, to ensure that drivers can see you.

7. Cycling Safety:

Always wear a helmet when riding a bicycle.

Use appropriate hand signals for turning and stopping.

Follow traffic rules and use bike lanes where available.

Ensure your bicycle is equipped with lights and reflectors if riding in dark or low-light conditions.

8. School Bus Safety:

Wait for the bus to come to a complete stop before approaching it.

Enter and exit the bus calmly and use the handrail.

After getting off the bus, wait for it to move away before crossing the road, ensuring it's safe.

9. Stay Alert at Bus Stops:

Wait at a safe distance from the road while waiting for the bus. Avoid pushing or crowding other students at the bus stop.

10. Never Run or Play Near Roads:

Avoid running or playing near roads or driveways. Always stay a safe distance from traffic.

11. Cross at Corners:

Cross the road at intersections or corners where vehicles are more likely to expect pedestrians.

12. Be Cautious Around Vehicles:

Never walk behind or in front of a parked vehicle without checking if the driver can see you and if the vehicle is about to move.

13. Use Seat Belts:

Always wear a seat belt when riding in a vehicle. If available, use a booster seat if it is recommended for your age and size.

14. Avoid Hitchhiking:

Never hitchhike or accept rides from strangers. Always use official school or public transport services.

15. Stay in Safe Zones:

If your school has a designated pick-up and drop-off zone, use it. Avoid crossing busy streets or intersections that are not supervised.

16. Learn Basic Traffic Signs:

Familiarize yourself with basic traffic signs and signals so you can understand and react to them appropriately.

17. Follow School Guidelines:

Adhere to any specific road safety guidelines provided by your school, such as safe routes or rules for using school transport.

Conclusion:

Implementing these road safety rules helps ensure that students are protected while travelling to and from school or other activities. Education on road safety should be complemented with practical experience and supervision to reinforce safe behaviour and decision-making on the roads.

3.5 Moral story on road safety

Title: The Road of Responsibility

Once upon a time, in a small village nestled between green hills, there was a young boy named Ravi. Ravi was known for his adventurous spirit. He loved riding his bicycle through the winding roads of the village, feeling the wind rush past him as he sped down the lanes.

One day, while riding his bicycle home from school, Ravi saw his friends riding fast and recklessly. They were performing stunts, racing with each other, and ignoring the road signs. Ravi, excited by the thrill, decided to join them. He started speeding down the road, ignoring the stop signs and swerving between pedestrians.

Suddenly, as Ravi turned a sharp corner, he lost control of his bicycle and crashed into a small cart filled with fruits. The cart toppled over, scattering fruits everywhere, and Ravi fell hard onto the ground. The elderly woman who owned the cart was shaken, but more than that, she was worried about Ravi.

Seeing what had happened, his friends quickly fled the scene, leaving Ravi alone. The old woman helped him up and said, "Son, are you alright?" Ravi, though bruised and ashamed, nodded.

She continued, "You must understand, the road is not a playground. It is a place where we must all be responsible, not just for ourselves, but for others too. Speeding and ignoring rules may seem fun, but it can lead to accidents that hurt people."

Ravi realized his mistake and apologized to the old woman. He helped her pick up the scattered fruits and offered to repair her cart with his father's help. From that day on, Ravi became a champion of road safety. He learned the importance of following traffic rules, riding his bicycle carefully, and always being mindful of others on the road.

He also encouraged his friends to do the same. Soon, the entire village noticed a change in the way the children rode their bicycles. They started wearing helmets, obeying traffic signs, and riding at safe speeds. The village became a safer place, and the elders were pleased to see the children setting an example for everyone.

The lesson Ravi learned stayed with him for life: "The road is a shared space, and with freedom comes responsibility. Being safe is being smart."

Moral of the story:

Road safety is not just about following rules; it's about caring for yourself and others. Always ride or drive responsibly, because one mistake can affect many lives.

3.6 Tips for Train Safety

Train safety is crucial for preventing accidents and ensuring a safe experience for passengers, pedestrians, and those working near railway tracks. Here are essential tips for train safety:

For Train Passengers:

Use Designated Areas:

Platform Safety: Always wait for trains in designated areas on the platform and behind the safety lines.

Boarding and Alighting: Wait for the train to come to a complete stop before boarding or alighting.

Stay Alert:

Avoid Distractions: Pay attention to announcements and visual information. Avoid using mobile phones or other distractions while waiting on the platform.

Keep Clear of the Tracks:

Track Safety: Never walk on or near the tracks, even if they appear to be clear. Always use designated pedestrian crossings.

Follow Instructions:

Safety Announcements: Follow all safety instructions and announcements made by train staff or through station PA systems.

Be Careful with Luggage:

Secure Luggage: Keep luggage and personal belongings clear of the platform edge and ensure they are stowed safely in the designated areas on the train.

Supervise Children:

Child Safety: Ensure children are supervised and remain behind the safety lines while waiting for the train. Hold their hand when boarding and alighting.

Emergency Procedures:

Know the Protocol: Familiarize yourself with emergency procedures on the train and at the station, including the location of emergency exits and equipment.

For Pedestrians Near Train Tracks:

Use Authorized Crossings:

Railroad Crossings: Always use marked pedestrian crossings or overpasses/underpasses when crossing tracks. Never attempt to cross where there is no designated crossing.

Obey Warning Signals:

Gates and Lights: Obey all warning signals at railroad crossings, including flashing lights, bells, and lowered gates.

Never Cross When Signals Are Active:

Active Warning Signals: Do not attempt to cross the tracks when warning signals are active, as a train may be approaching.

Stay Clear of Trains:

Safe Distance: Maintain a safe distance from the edge of the tracks, and never attempt to cross in front of a moving train.

Be Cautious at Night:

Visibility: Be extra cautious at night or in poor visibility conditions. Ensure you can see and hear oncoming trains.

Avoid Using Headphones:

Hearing Awareness: Do not use headphones or earbuds while crossing tracks, as you need to be able to hear approaching trains.

For Drivers at Railroad Crossings:

Stop at Crossings:

Crossing Gates: Stop when the crossing gates are down, or when signals are flashing. Proceed only when it is safe and the gates are fully raised.

Check for Multiple Trains:

Look and Listen: Always check for additional trains when crossing tracks, as trains may follow each other in quick succession.

Never Stop on Tracks:

Clear the Tracks: Ensure your vehicle clears the tracks completely before stopping. Do not stop or park on the tracks.

Follow the Signals:

Signal Compliance: Obey all warning signals and signage at railroad crossings. Do not attempt to beat a train by speeding up.

Be Aware of Train Speeds:

Speed Awareness: Trains can approach quickly and may be difficult to judge in terms of speed and distance. Be cautious and patient.

For Workers and Maintenance Personnel:

Wear Safety Gear:

Protective Equipment: Always wear appropriate safety gear, such as high-visibility clothing, hard hats, and safety boots.

Follow Safety Protocols:

Workplace Safety: Adhere to all safety protocols and procedures when working near or on railway tracks. Use warning signals and barriers as required.

Communicate Clearly:

Team Coordination: Maintain clear communication with other team members and train operators when working in or near railway areas.

Be Aware of Train Schedules:

Schedule Awareness: Be informed about train schedules and movements to avoid working in areas where trains are expected.

Conclusion:

Following these train safety tips helps prevent accidents and ensures a safer environment for everyone involved with rail travel and work. Awareness, vigilance, and adherence to safety protocols are key to maintaining safety around trains and railway tracks.

3.7 Personal experience in Road safety

Here's an example of a personal experience shared by a car driver on road safety:

Title: A Lesson in Road Safety: My Close Call on a Rainy Day

I've always considered myself a cautious driver. I follow speed limits, wear my seatbelt, and avoid distractions while driving. But one rainy day, I learned that even the smallest mistake on the road could lead to a potentially dangerous situation.

It was a typical afternoon commute. The rain had been coming down steadily all day, making the roads slick. I was driving on a familiar highway, heading home from work. As usual, I kept my speed in check, knowing that wet roads could be tricky.

As I approached a sharp curve, I noticed the car in front of me start to skid. The driver overcorrected, and suddenly, the car spun out of control. My heart raced as I realized I was too close to avoid the car if it completely spun out. I instinctively pressed on the brakes, but with the wet road, my car started to slide as well.

Fortunately, I had recently read about the importance of maintaining a safe following distance, especially in bad weather. Because of this, I had enough space to slow down without crashing into the spinning car. The other driver managed to regain control, and we both came to a stop safely.

That experience was a wake-up call. It made me realize how critical it is to stay alert and prepared for the unexpected, especially in challenging conditions like rain. Now, I make it a point to increase my following distance even more in bad weather and stay extra vigilant. That close call could have been much worse, but it taught me an invaluable lesson about the importance of road safety.

This experience highlights the importance of adjusting driving behaviour according to road conditions and the value of maintaining a safe following distance.

3.8 Brainstorm

Here is a 30-question multiple-choice test on road safety:

Road Safety Multiple-Choice Test

1. What is the primary purpose of road signs?

   a) To decorate the road

   b) To provide information and ensure safety

   c) To confuse drivers

   d) To promote businesses

2. What should you do when you see a yellow traffic light?

   a) Speed up to cross before it turns red

   b) Stop immediately

   c) Prepare to stop if it is safe to do so

   d) Ignore it and keep driving

3. Which of the following is considered a distraction while driving?

   a) Adjusting mirrors before driving

   b) Talking on a hands-free device

   c) Eating while driving

   d) Both b and c

4. What is the legal blood alcohol concentration (BAC) limit for drivers in most countries?

a) 0.05%

b) 0.10%

c) 0.08%

d) 0.12%

5. Which of the following actions is the best way to avoid fatigue while driving?

   a) Drinking coffee

   b) Opening the window for fresh air

   c) Taking regular breaks and resting

   d) Turning up the music volume

6. What does a flashing red traffic light mean?

   a) Stop, and then proceed when safe

   b) Slow down

   c) Yield to oncoming traffic

   d) The light is malfunctioning

7. When should you use your seatbelt?

   a) Only on highways

   b) Only when driving long distances

   c) Every time you are in a moving vehicle

   d) Only when the driver instructs you to do so

8. What is the recommended following distance under normal driving conditions?

a) One second

b) Two seconds

c) Three seconds

d) Four seconds

9. Which of the following is the most common cause of road accidents?

   a) Mechanical failure

   b) Adverse weather conditions

   c) Driver error

   d) Poor road design

10. In the event of a tyre blowout, what is the best course of action?

    a) Brake immediately and pull over

    b) Keep a firm grip on the steering wheel and slowly reduce speed

    c) Turn the wheel sharply to regain control

    d) Speed up to stabilize the vehicle

11. What should you do if your vehicle starts to skid?

    a) Slam on the brakes

    b) Steer in the direction you want the vehicle to go

    c) Turn the wheel in the opposite direction of the skid

    d) Accelerate to regain control

12. When is it safe to overtake another vehicle?

a) When approaching a curve

b) When the road ahead is clear and there are no oncoming vehicles

c) When the vehicle in front is driving slowly, regardless of conditions

d) When you are in a no-overtaking zone

13. What is the purpose of a pedestrian crossing?

a) To allow pedestrians to cross the road safely

b) To slow down traffic

c) To provide a place for vehicles to stop

d) To decorate the road

14. Which of the following is true about using a mobile phone while driving?

a) It is safe as long as you use a hands-free device

b) It is safer to text at red lights only

c) It is a major distraction and should be avoided

d) It helps to keep drivers alert

15. What does a double solid yellow line on the road indicate?

a) Passing is allowed

b) Passing is prohibited in both directions

c) Passing is allowed in one direction only

d) The road is under construction

16. When driving in foggy conditions, you should:

a) Use high-beam headlights

b) Use low-beam headlights or fog lights

c) Drive as fast as possible to get out of the fog

d) Turn off your headlights

17. What should you do when approaching a school bus that has stopped and is displaying red flashing lights?

a) Slow down and pass carefully

b) Stop, regardless of your direction of travel

c) Honk to alert the driver

d) Overtake quickly before the children cross

18. What is the minimum safe distance to stop behind another vehicle at a traffic light?

a) One car length

b) Two car lengths

c) Enough space to see the rear tyres of the vehicle in front touching the ground

d) No minimum distance is required

19. If you are involved in a minor accident and no one is injured, you should:

a) Leave the scene immediately

b) Move your vehicle off the road, if possible, to avoid blocking traffic

c) Argue with the other driver about who was at fault

d) Call your insurance company on the spot

20. When should you yield the right of way?

    a) Only at stop signs

    b) When entering a roundabout

    c) When you are in a hurry

    d) Only when the other driver honks

21. Which of the following is the safest way to deal with aggressive drivers?

    a) Speed up to get away from them

    b) Engage in a confrontation

    c) Stay calm and avoid eye contact

    d) Brake suddenly to make them back off

22. What does a red triangular yield sign mean?

    a) Stop

    b) Yield to oncoming traffic

    c) No entry

    d) Give way to all traffic

23. When should you use your horn?

    a) When angry with another driver

    b) To alert other drivers of your presence

    c) To express frustration

d) Continuously to clear the road

24. What is the safest way to change lanes?

    a) Change lanes quickly without signalling

    b) Check your mirrors and blind spots, signal, and then change lanes smoothly

    c) Speed up and change lanes abruptly

    d) Signal after you have started changing lanes

25. When driving through a construction zone, you should:

    a) Ignore all signs and follow your usual speed

    b) Reduce your speed and follow all signs and instructions

    c) Overtake construction vehicles

    d) Drive at the same speed as before

26. What should you do if you approach an intersection with a malfunctioning traffic light?

    a) Proceed as if the light is green

    b) Stop and proceed only when it is safe

    c) Speed up to clear the intersection quickly

    d) Follow the vehicle in front of you without stopping

27. Which of the following factors increases the risk of hydroplaning?

    a) Dry road conditions

    b) Driving at high speeds on wet roads

c) Having properly inflated tyres

d) Slowing down when driving through water

28. What is the best way to handle a situation where you encounter a wrong-way driver on the road?

    a) Honk your horn and flash your lights

    b) Try to overtake them

    c) Move to the shoulder and stop, if possible

    d) Continue driving in your lane

29. What should you do if an emergency vehicle with flashing lights is approaching from behind?

    a) Speed up to get out of the way

    b) Pull over to the right and stop

    c) Stop immediately in your lane

    d) Ignore it and keep driving

30. When parking on a hill facing uphill with a curb, you should:

    a) Turn your wheels toward the curb

    b) Turn your wheels away from the curb

    c) Leave your wheels straight

    d) Engage the handbrake only

Answers:

1. b) To provide information and ensure safety

2. c) Prepare to stop if it is safe to do so
3. d) Both b and c
4. c) 0.08%
5. c) Taking regular breaks and resting
6. a) Stop, and then proceed when safe
7. c) Every time you are in a moving vehicle
8. c) Three seconds
9. c) Driver error
10. b) Keep a firm grip on the steering wheel and slowly reduce speed
11. b) Steer in the direction you want the vehicle to go
12. b) When the road ahead is clear and there are no oncoming vehicles
13. a) To allow pedestrians to cross the road safely
14. c) It is a major distraction and should be avoided
15. b) Passing is prohibited in both directions
16. b) Use low-beam headlights or fog lights
17. b) Stop, regardless of your direction of travel
18. c) Enough space to see the rear tyres of the vehicle in front touching the ground
19. b) Move your vehicle off the road, if possible, to avoid blocking traffic

20. b) When entering a roundabout

21. c) Stay calm and avoid eye contact

22. d) Give way to all traffic

23. b) To alert other drivers of your presence

24. b) Check your mirrors and blind

25. b) Reduce your speed and follow all signs and instructions

26. b) Stop and proceed only when it is safe

27. d) Slowing down when driving through water

28/ a) Honk your horn and flash your lights

29. a) Speed up to get out of the way

30. d) Engage the handbrake only

Part 4

Fire Safety

PART IV

4-FIRE SAFETY

4.1 Introduction to fire safety

Fire safety is a critical aspect of public and personal safety, encompassing practices and measures designed to prevent, detect, and respond to fires effectively. Fire incidents can have devastating consequences, leading to property damage, injury, and loss of life. Therefore, understanding and implementing fire safety protocols is essential in mitigating these risks and ensuring a safe environment.

Introduction to Fire Safety

Fire safety involves a range of strategies and actions aimed at preventing the outbreak of fires, minimizing their impact, and protecting people and property in the event of a fire. It encompasses both proactive measures to reduce the likelihood of a fire occurring and reactive measures to manage the situation if a fire does start.

Key Components of Fire Safety:

Prevention:

Fire Risk Assessment: Identifying potential fire hazards and implementing measures to eliminate or control these risks.

Safe Practices: Adopting safe practices in both residential and commercial settings, such as proper storage of flammable materials and regular maintenance of electrical systems.

Detection:

Smoke Detectors: Installing smoke alarms and detectors that alert occupants to the presence of smoke or fire.

Fire Alarms: Utilizing fire alarm systems in commercial buildings to provide early warning and facilitate evacuation.

Response:

Fire Extinguishers: Providing and maintaining fire extinguishers for use in the early stages of a fire to control small fires.

Emergency Plans: Developing and practising emergency response plans, including evacuation procedures and communication strategies.

Education:

Training: Educating individuals and organizations about fire safety practices, including how to use fire extinguishers, conduct fire drills, and understand fire alarms.

Public Awareness: Raising awareness about fire safety through campaigns and educational programs.

Importance of Fire Safety:

Protection of Lives: The primary goal of fire safety is to protect the lives of individuals by ensuring that they have the knowledge and tools to respond effectively to a fire emergency.

Minimization of Damage: Implementing fire safety measures helps reduce property damage and financial loss caused by fires.

Regulatory Compliance: Adhering to fire safety regulations and standards is required by law in many regions and ensures compliance with safety codes.

Common Fire Safety Practices:

Regular Inspections: Conduct regular inspections of fire safety equipment and systems to ensure they are functioning properly.

Emergency Exits: Marking and keeping emergency exits accessible to facilitate quick evacuation.

Fire Drills: Conducting regular fire drills to familiarize occupants with evacuation routes and procedures.

In summary, fire safety is an essential aspect of protecting people and property from the dangers of fire. By implementing preventive measures, ensuring effective detection and response systems, and promoting education and awareness, individuals and organizations can significantly reduce the risk of fire-related incidents and enhance overall safety.

4.2 The Fire Triangle

The Fire Triangle is a fundamental concept in fire safety that explains the three essential elements required for a fire to ignite and sustain itself. Understanding the Fire Triangle is crucial for both preventing and extinguishing fires. The three elements of the Fire Triangle are:

1. Heat:

Description: Heat is the energy source that raises the temperature of a material to its ignition point. It can come from various sources, such as open flames, electrical equipment, machinery, and friction.

Examples: Matches, electrical sparks, hot surfaces, and heating appliances.

Fire Prevention Tip: Control and manage heat sources to prevent them from coming into contact with flammable materials. Regularly inspect and maintain heating systems and electrical equipment.

2. Fuel:

Description: Fuel is any combustible material that can be ignited by heat. It provides the substance that burns and sustains the fire.

Examples: Wood, paper, gasoline, oils, textiles, and other flammable materials.

Fire Prevention Tip: Store fuels safely in designated areas and away from heat sources. Keep combustible materials properly managed and minimize their accumulation in vulnerable areas.

3. Oxygen:

Description: Oxygen is the element that supports the chemical reaction of combustion. It is necessary for the fuel to burn and for the fire to continue.

Examples: Ambient air (which contains about 21% oxygen) and pure oxygen supplies.

Fire Prevention Tip: Ensure good ventilation to prevent the accumulation of flammable gases or vapours. Avoid creating conditions where oxygen can increase fire intensity, such as in confined spaces with inadequate ventilation.

Fire Triangle and Fire Extinguishment:

To extinguish a fire, at least one of the three elements of the Fire Triangle must be removed or controlled. This is the basis for various fire suppression methods:

Remove Heat: Use water or other cooling agents to lower the temperature of the fire and reduce its heat. This method is effective for fires involving common combustibles.

Remove Fuel: Remove or isolate the fuel source from the fire. This can be done by moving combustible materials away or shutting off the supply of flammable liquids or gases.

Remove Oxygen: Use fire suppressants like foam, CO2, or dry chemical agents to displace or reduce the oxygen available to the fire. This method is effective for fires involving flammable liquids and electrical equipment.

Conclusion:

The Fire Triangle is a fundamental concept that highlights the three critical components required for a fire to start and sustain. By understanding and managing these elements, individuals and organizations can implement effective fire prevention strategies and extinguish fires more efficiently. Removing one or more of the elements can help control and extinguish fires, ensuring better safety and reducing fire risks.

4.3 The types of fire extinguishers

Fire extinguishers are classified based on the type of fire they are designed to combat. Understanding the different types of fire extinguishers and their applications is crucial for effective fire safety. Here's a detailed overview of the various types of fire extinguishers:

1. Water Fire Extinguishers (Class A):

Description: Water fire extinguishers are used primarily for fires involving ordinary combustible materials such as wood, paper, cloth, and some plastics.

How It Works: Water cools the burning material below its ignition temperature and removes heat from the fire.

Identification: Typically, red and marked with a white "A" symbol.

Usage: Not suitable for electrical fires, flammable liquids, or cooking oils.

2. Foam Fire Extinguishers (Class A and B):

Description: Foam fire extinguishers are effective on both Class A (combustibles) and Class B (flammable liquids) fires.

How It Works: Foam creates a barrier that smothers the fire, cooling the burning material and preventing the release of flammable vapours.

Identification: Usually red with a cream or beige label marked with "A" and "B" symbols.

Usage: Ideal for fires involving flammable liquids and solids but not suitable for electrical fires.

3. Carbon Dioxide (CO_2) Fire Extinguishers (Class B and Electrical):

Description: CO_2 fire extinguishers are designed for Class B (flammable liquids) and electrical fires.

How It Works: CO_2 displaces oxygen and cools the fire, effectively suffocating the flames.

Identification: Typically, red with a black label marked with "B" and a symbol for electrical fires.

Usage: Effective on electrical fires and flammable liquids but not suitable for Class A fires or cooking oils.

4. Dry Chemical Powder Fire Extinguishers (Class A, B, and C):

Description: Dry chemical powder extinguishers are versatile and can be used on Class A (combustibles), Class B (flammable liquids), and Class C (electrical) fires.

How It Works: The powder interrupts the chemical reaction of the fire, smothering and extinguishing the flames.

Identification: Usually red with a blue label marked with "A," "B," and "C" symbols.

5. Clean Agent Fire Extinguishers (Class A, B, and C):

Description: Clean agent extinguishers use gases like Halon or alternatives (e.g., FM-200, FE-36) that leave no residue.

How It Works: These agents interrupt the fire's chemical reaction and are effective on Class A, B, and C fires.

Identification: Usually red with a label indicating the type of agent used.

Usage: Ideal for sensitive areas such as data centres and laboratories where residue or cleanup could be a problem.

6. Water Mist Fire Extinguishers (Class A and Electrical):

Description: Water mist extinguishers use fine mist to extinguish fires.

How It Works: The mist cools the fire and reduces the risk of re-ignition while minimizing water damage.

Identification: Usually red with a blue label.

Usage: Effective on Class A and electrical fires; not suitable for flammable liquids.

Choosing the Right Fire Extinguisher:

Class A Fires: Use water, foam, or dry chemical powder extinguishers.

Class B Fires: Use foam, CO2, dry chemical powder, or clean agents.

Class C Fires: Use CO2, dry chemical powder, or clean agents.

Class D Fires: Use wet chemical extinguishers.

4.4 How to Use a Fire Extinguisher

Using a fire extinguisher effectively involves understanding the correct technique and applying it properly during a fire emergency. The most widely used method for operating a fire extinguisher is the PASS technique. Here's a step-by-step guide on how to use a fire extinguisher:

PASS Technique

Pull the Pin:

Action: Pull the safety pin or locking mechanism from the handle of the fire extinguisher. This pin prevents accidental discharge.

Purpose: Removing the pin allows you to operate the extinguisher.

Aim the Nozzle:

Action: Stand a safe distance away from the fire (typically 6 to 8 feet) and aim the nozzle or hose at the base of the fire.

Purpose: Targeting the base of the fire ensures you're attacking the fuel source rather than the flames.

Squeeze the Handle:

Action: Squeeze the handle or lever to release the extinguishing agent.

Purpose: Discharging the agent helps to put out the fire.

Sweep the Nozzle:

Action: Sweep the nozzle or hose from side to side while continuing to aim at the base of the fire.

Purpose: This ensures even coverage and helps to extinguish the fire completely.

Additional Tips for Using a Fire Extinguisher:

Assess the Fire: Ensure the fire is small and manageable. If the fire is large, immediately evacuate and call emergency services. Fire extinguishers are for small, contained fires.

Maintain a Safe Distance: Keep a safe distance from the fire, and approach it with caution. Be aware of any changes in the fire's behaviour.

Stand Back: If the fire starts to spread or if you are unable to extinguish it, retreat to a safe location and evacuate the area.

Know Your Extinguisher: Familiarize yourself with the type of extinguisher and its intended use. Different extinguishers are designed for different classes of fires.

Read the Instructions: Before using an extinguisher, read the instructions on the label to understand how it operates.

When to Use a Fire Extinguisher:

Class A Fires: Involving ordinary combustibles like wood, paper, or cloth.

Class B Fires: Involving flammable liquids such as gasoline, oil, or grease.

Class C Fires: Involving electrical equipment, provided the power source has been turned off.

Class D Fires: Burning materials.

When Not to Use a Fire Extinguisher:

Large Fires: If the fire is too large to handle or is spreading rapidly, evacuate the area and call the fire department immediately.

Unfamiliar Fires: If you are unsure of the type of fire or how to use the extinguisher properly, do not attempt to fight the fire.

Blocked Exits: Ensure that you have a clear escape route before attempting to use the extinguisher.

Post-Use Actions:

Report the Incident: Notify the appropriate personnel (e.g., building manager or safety officer) of the fire incident and any extinguisher usage.

Replace or Recharge: After using an extinguisher, have it inspected, recharged, or replaced as needed.

Check for Damage: Inspect the area for any remaining fire hazards or damage.

Conclusion:

The PASS technique is a simple and effective method for using a fire extinguisher. Proper training and understanding of how to operate different types of extinguishers are essential for effective fire management and safety. Always prioritize safety and evacuate if the fire becomes uncontrollable or if you are unsure about handling it.

4.5 Success story

The Kumbakonam School Fire Tragedy and the Lessons Learned

Background

On July 16, 2004, a devastating fire broke out at the Sri Krishna Aided Primary School in Kumbakonam, Tamil Nadu, India. The fire began in the kitchen, where a meal was being prepared for the students. The school, which housed over 200 students, was constructed with thatched roofs made of coconut leaves—a highly flammable material. The fire quickly spread to the classrooms, trapping many young students inside.

The Tragedy

The fire claimed the lives of 94 children, all between the ages of 6 and 10, making it one of the deadliest school fires in Indian history. The tragedy was a result of several fire safety lapses, including inadequate emergency exits, poor building materials, and a lack of fire safety training among the staff.

The Response and Reforms

In the aftermath of the Kumbakonam fire tragedy, the entire nation was shocked, and there was an urgent call for reform in fire safety regulations, especially in schools.

1. Increased Awareness: The tragedy led to increased awareness about the importance of fire safety in educational institutions across India. Authorities began inspecting schools to ensure they complied with fire safety norms.

2. Fire Safety Audits: The Tamil Nadu government mandated fire safety audits for all schools in the state. This included ensuring proper exits, the installation of fire extinguishers, and the removal of flammable materials from school premises.
3. Training and Drills: Schools were required to conduct regular fire drills to train students and staff on how to respond in the event of a fire. This was a significant change, as many schools previously did not have any formal fire safety procedures in place.
4. Building Code Revisions: The fire also prompted revisions to the National Building Code of India. The code was updated to include stricter regulations on the materials used in school buildings, ensuring they were non-flammable and met safety standards.

A Positive Outcome

One of the most remarkable success stories that emerged from this tragedy was the transformation in fire safety culture across Tamil Nadu and, subsequently, other parts of India. Schools that once neglected fire safety now took it seriously. In the years following the Kumbakonam fire, there were numerous instances where schools successfully evacuated students during fires, avoiding what could have been catastrophic incidents.

For example, in 2008, a fire broke out in a school in the Thanjavur district of Tamil Nadu. Thanks to the fire safety drills and the newly implemented safety measures, the students and staff were able to evacuate the building quickly and safely, with no casualties.

This incident highlighted the effectiveness of the reforms that were put in place after the Kumbakonam tragedy.

The Legacy

The Kumbakonam fire tragedy remains a painful memory in India's history, but it also stands as a turning point in how the country approaches fire safety in schools. The reforms and awareness that followed have undoubtedly saved many lives, proving that even in the wake of tragedy, positive change is possible.

The legacy of those 94 children lives on in the improved safety standards that now protect millions of students across India, ensuring that such a tragedy never happens again.

4.6 Brainstorm

Here's a 40-question multiple-choice test on fire safety:

Fire Safety Multiple-Choice Test

1. What is the most common source of home fires?

   a) Faulty electrical wiring

   b) Cooking equipment

   c) Candles

   d) Portable heaters

2. Which type of fire extinguisher should be used on a grease fire?

   a) Water

   b) $CO2$

   c) Dry chemical

   d) Foam

3. What is the first thing you should do if you smell smoke in your home?

   a) Investigate the source

   b) Open windows

   c) Get everyone out and call 911

   d) Use a fan to clear the smoke

4. Which of the following is a key fire prevention strategy?

   a) Keeping exits locked

   b) Storing flammable materials near heat sources

   c) Regularly testing smoke alarms

   d) Overloading electrical outlets

5. Why is it important to have a fire escape plan?

   a) To avoid panic during a fire

   b) To ensure all doors are locked

   c) To keep pets safe

   d) To reduce the likelihood of a fire occurring

6. Which of the following is NOT a class of fire?

   a) Class A: Ordinary combustibles

   b) Class B: Flammable liquids

   c) Class C: Electrical

   d) Class D: Radiation

7. What should you do if your fire extinguisher is not working?

   a) Continue trying to use it

   b) Leave it and evacuate

   c) Refill it with water

   d) Call a professional to fix it

8. Where should smoke alarms be installed in a home?

   a) In every bedroom

b) In the kitchen only

c) Near electrical panels

d) In the basement only

9. What is the purpose of a fire drill?

   a) To test fire alarms

   b) To practice evacuation procedures

   c) To locate fire hazards

   d) To meet fire safety officials

10. What should you do if your clothes catch fire?

    a) Run for help

    b) Stop, drop, and roll

    c) Use water to douse the flames

    d) Tear off the burning clothes

11. What type of fire is a Class B fire extinguisher designed for?

    a) Electrical fires

    b) Cooking oil fires

    c) Flammable liquid fires

    d) Wood and paper fires

12. How often should fire escape plans be practised?

    a) Once a year

    b) Twice a year

c) Monthly

d) Every other year

13. What is the recommended distance to stay away from a fire when using a fire extinguisher?

a) 1 foot

b) 5 feet

c) 8 feet

d) 12 feet

14. What is the first step you should take before using a fire extinguisher?

a) Shout for help

b) Read the instructions

c) Check the type of fire

d) Pull the pin

15. What should you do if a door feels hot during a fire?

a) Open it slowly

b) Force it open

c) Use another exit

d) Wait for it to cool down

16. Why should you never use an elevator during a fire evacuation?

a) It may malfunction and trap you

b) It's faster to take the stairs

c) The elevator is reserved for firefighters

d) Smoke rises faster in elevators

17. How should you store flammable liquids at home?

    a) In the kitchen

    b) In a cool, dry place away from heat

    c) In the garage

    d) Near electrical outlets

18. What is a fire break?

    a) A tool used to fight fires

    b) A cleared area to stop the spread of fire

    c) A device that prevents fire from starting

    d) A scheduled fire drill

19. Which of these materials is most fire-resistant?

    a) Cotton

    b) Polyester

    c) Wool

    d) Nylon

20. What should you do if you are trapped in a room during a fire?

    a) Break the windows

    b) Seal the door gaps with wet cloths

    c) Hide under a bed

d) Try to fight the fire

21. Which fire class is associated with electrical fires?

    a) Class A

    b) Class B

    c) Class C

    d) Class D

22. What does the acronym PASS stand for when using a fire extinguisher?

    a) Pull, Aim, Squeeze, Sweep

    b) Press, Aim, Squeeze, Spray

    c) Push, Activate, Spray, Sweep

    d) Pull, Activate, Squeeze, Sweep

23. What should be the first priority during a fire emergency?

    a) Saving belongings

    b) Extinguishing the fire

    c) Evacuating people safely

    d) Calling the fire department

24. How often should you replace the batteries in a smoke detector?

    a) Every month

    b) Every six months

    c) Once a year

d) Every five years

25. Why is it important to close doors behind you during a fire evacuation?

    a) To contain the fire

    b) To prevent drafts

    c) To keep pets inside

    d) To reduce noise

26. What is a fire retardant?

    a) A substance that fuels a fire

    b) A material that resists burning

    c) A tool for starting a fire

    d) A type of fire extinguisher

27. What type of smoke alarm is best for detecting slow, smouldering fires?

    a) Ionization smoke alarm

    b) Photoelectric smoke alarm

    c) Dual-sensor smoke alarm

    d) Carbon monoxide detector

28. Which action is unsafe when using a space heater?

    a) Keeping it on a flat surface

    b) Plugging it into an extension cord

    c) Turning it off when not in use

d) Keeping flammable materials away

29. What does a carbon monoxide detector do?

    a) Detects fires

    b) Detects smoke

    c) Detects flammable gases

    d) Detects carbon monoxide levels

30. What is the safest way to evacuate a burning building?

    a) Use the elevator

    b) Run as fast as you can

    c) Walk calmly and follow the exit signs

    d) Jump out of a window.

Answers:

1. b) Cooking equipment
2. c) Dry chemical
3. c) Get everyone out and call 911
4. c) Regularly testing smoke alarms
5. a) To avoid panic during a fire
6. d) Class D: Radiation
7. b) Leave it and evacuate
8. a) In every bedroom
9. b) To practice evacuation procedures

10. b) Stop, drop, and roll
11. c) Flammable liquid fires
12. b) Twice a year
13. c) 8 feet
14. d) Pull the pin
15. c) Use another exit
16. a) It may malfunction and trap you
17. b) In a cool, dry place away from heat
18. b) A cleared area to stop the spread of fire
19. c) Wool
20. b) Seal the door gaps with wet cloths
21. c) Class C
22. a) Pull, Aim, Squeeze, Sweep
23. c) Evacuating people safely
24. b) Every six months
25. a) To contain the fire
26. b) A material that resists burning
27. b) Photoelectric smoke alarm
28. b) Plugging it into an extension cord
29. d) Detects carbon monoxide levels
30. c) Walk calmly and follow the exit signs

Part 5

Environment Safety

PART V

5.-ENVIRONMENT SAFETY

5.1 All about Environmental Safety

Environmental safety encompasses the practices, policies, and procedures aimed at protecting the environment from harmful human activities and ensuring the safety of both ecosystems and human health. It involves the responsible management of natural resources, pollution prevention, and minimizing the impact of industrial and agricultural activities on the environment. Here's an in-depth look at the various aspects of environmental safety:

1. Understanding Environmental Safety:

Definition: Environmental safety refers to the safeguarding of the environment from potential harm caused by human activities, including industrial processes, waste disposal, chemical usage, and natural resource exploitation.

Goal: The primary goal of environmental safety is to reduce the risk of environmental degradation, protect natural resources, and promote sustainable development.

2. Key Components of Environmental Safety:

a. Pollution Prevention:

Air Pollution: Implementing measures to reduce emissions of

harmful gases, particulates, and volatile organic compounds (VOCs) from industrial activities, transportation, and energy production.

Water Pollution: Preventing contamination of water bodies through proper waste disposal, treatment of industrial effluents, and minimizing agricultural runoff.

Soil Pollution: Reducing soil contamination by managing the use of pesticides, fertilizers, and industrial chemicals, and promoting safe disposal of hazardous wastes.

b. Waste Management:

Solid Waste: Implementing effective waste management strategies, including recycling, composting, and safe disposal of non-recyclable materials.

Hazardous Waste: Proper handling, storage, and disposal of hazardous materials to prevent environmental contamination and health risks.

E-waste: Managing electronic waste through recycling programs and safe disposal to reduce the release of toxic substances.

c. Resource Conservation:

Energy Conservation: Promoting the use of energy-efficient technologies and renewable energy sources to reduce the environmental impact of energy consumption.

Water Conservation: Implementing water-saving practices, such as rainwater harvesting, water recycling, and efficient irrigation methods.

Sustainable Agriculture: Encouraging sustainable farming practices that minimize the use of chemical inputs, promote soil health, and protect biodiversity.

d. Environmental Impact Assessment (EIA):

Purpose: Conducting EIAs to evaluate the potential environmental effects of proposed projects or developments before they are undertaken.

Process: Assessing factors such as air and water quality, biodiversity, and social impact to identify mitigation measures and alternatives.

e. Climate Change Mitigation:

Greenhouse Gas Reduction: Implementing strategies to reduce emissions of greenhouse gases (GHGs) through energy efficiency, renewable energy, and carbon capture.

Adaptation Measures: Develop adaptive strategies to cope with the impacts of climate change, such as rising sea levels, extreme weather events, and changing ecosystems.

f. Environmental Regulations and Compliance:

Legislation: Adhering to environmental laws and regulations at the local, national, and international levels to ensure compliance with environmental safety standards.

Monitoring and Reporting: Regular monitoring of environmental parameters and reporting to regulatory authorities to ensure ongoing compliance and transparency.

3. Environmental Safety in Industry:

a. Sustainable Manufacturing:

Eco-friendly Processes: Implementing manufacturing processes that reduce waste, minimize energy consumption, and utilize renewable resources.

Product Lifecycle Management: Designing products with their entire lifecycle in mind, from raw material extraction to disposal, to minimize environmental impact.

b. Chemical Safety:

Safe Handling: Ensuring the safe handling, storage, and disposal of chemicals to prevent spills, leaks, and contamination.

Substitution: Replacing hazardous chemicals with safer alternatives where possible to reduce environmental and health risks.

c. Workplace Environmental Safety:

Employee Training: Providing training to employees on environmental safety practices, such as spill response, waste management, and energy conservation.

Environmental Management Systems (EMS): Implementing EMS frameworks, such as ISO 14001, to systematically manage and improve environmental performance.

4. Environmental Safety in the Community:

a. Public Awareness:

Education Campaigns: Raising public awareness about environmental issues and promoting eco-friendly behaviours, such as recycling, energy conservation, and reducing single-use plastics.

Community Involvement: Engaging communities in environmental conservation projects, such as tree planting, clean-up drives, and water conservation initiatives.

b. Sustainable Urban Planning:

Green Spaces: Incorporating parks, gardens, and natural reserves in urban planning to enhance air quality, biodiversity, and public well-being.

Eco-friendly Transportation: Promoting the use of public transportation, cycling, and walking to reduce vehicle emissions and traffic congestion.

5. Challenges in Environmental Safety:

Resource Depletion: Overexploitation of natural resources, such as forests, water, and fossil fuels, leads to resource scarcity and environmental degradation.

Pollution Control: Managing pollution from industrial, agricultural, and urban sources remains a significant challenge, particularly in rapidly developing regions.

Climate Change: Addressing the global threat of climate change requires coordinated international efforts and significant reductions in GHG emissions.

Biodiversity Loss: Habitat destruction, pollution, and climate change are leading to the loss of biodiversity, which threatens ecosystem stability and resilience.

6. Best Practices for Enhancing Environmental Safety:

Adopt Sustainable Practices: Encourage the use of renewable energy, sustainable agriculture, and eco-friendly products in daily activities and business operations.

Implement Environmental Management Systems: Use frameworks

like ISO 14001 to systematically manage environmental impact and improve performance.

Engage in Continuous Improvement: Regularly review and update environmental safety policies and practices to adapt to new challenges and technologies.

Promote Collaboration: Work with government agencies, NGOs, and communities to develop and implement effective environmental safety strategies.

5.2 Environmental Safety Plan (ESP)

An Environmental Safety Plan (ESP) is a comprehensive document that outlines the policies, procedures, and practices an organization will implement to ensure the safety of the environment, employees, and the surrounding community from potential environmental hazards. It serves as a strategic guide for managing environmental risks, complying with regulations, and promoting sustainable practices. The plan is tailored to the specific needs of an organization and its operations, and it is regularly updated to reflect changes in regulations, technologies, and organizational goals.

Key Components of an Environmental Safety Plan:

Introduction and Objectives:

Purpose: Defines the purpose of the Environmental Safety Plan, including the organization's commitment to environmental protection and sustainability.

Objectives: Outlines specific goals, such as minimizing environmental impact, ensuring compliance with environmental laws, and promoting safe practices within the organization.

Regulatory Compliance:

Legal Requirements: Identifies relevant environmental laws, regulations, and standards that the organization must comply with, such as local, state, and federal environmental regulations.

Permits and Licenses: Details the required environmental permits and licenses, including procedures for obtaining and maintaining them.

Environmental Risk Assessment:

Hazard Identification: Identifies potential environmental hazards related to the organization's activities, such as chemical spills, emissions, waste generation, and resource depletion.

Risk Analysis: Assesses the likelihood and impact of identified hazards, prioritizing them based on the level of risk they pose to the environment and human health.

Pollution Prevention and Control Measures:

Air Quality Management: Outlines strategies to reduce air emissions, such as using cleaner fuels, optimizing processes, and installing pollution control equipment.

Water Quality Management: Includes measures to prevent water pollution, such as treating wastewater, reducing runoff, and protecting water sources.

Waste Management: Describes procedures for reducing, reusing, recycling, and safely disposing of waste materials, including hazardous waste.

Emergency Preparedness and Response:

Emergency Plans: Details procedures for responding to environmental emergencies, such as chemical spills, fires, and natural disasters.

Response Teams: Identifies key personnel responsible for emergency response, including their roles, responsibilities, and training requirements.

Communication: Establishes protocols for internal and external communication during an emergency, including notifying regulatory agencies and the public.

Training and Awareness:

Employee Training: Specifies training programs for employees to ensure they understand environmental safety procedures, risk management, and emergency response.

Public Awareness: Includes initiatives to raise awareness among the community and stakeholders about the organization's environmental safety efforts.

Environmental Monitoring and Reporting:

Monitoring Programs: Outlines procedures for regular monitoring of environmental parameters, such as air and water quality, waste levels, and energy usage.

Data Collection: Describes methods for collecting and analysing environmental data to track performance against objectives.

Reporting: Establishes reporting protocols for internal reviews and external communication with regulatory bodies, stakeholders, and the public.

Sustainable Practices and Resource Management:

Resource Conservation: Details strategies for conserving energy, water, and other natural resources, including the use of renewable energy sources and efficient technologies.

Sustainable Procurement: Encourages the use of environmentally friendly products and services, including suppliers that meet environmental standards.

Biodiversity Protection: Includes measures to protect local ecosystems and biodiversity, such as habitat conservation and reducing the impact of operations on wildlife.

Audits and Continuous Improvement:

Environmental Audits: Describes the process for conducting regular environmental audits to assess compliance, identify areas for improvement, and ensure the effectiveness of the ESP.

Corrective Actions: Outlines procedures for addressing non-compliance and implementing corrective actions to prevent future occurrences.

Plan Review and Update: Establishes a schedule for reviewing and updating the ESP to reflect changes in regulations, technology, and organizational goals.

Documentation and Record Keeping:

Record Maintenance: Specifies the types of records to be maintained, such as permits, training logs, monitoring data, and incident reports.

Documentation Procedures: Describes how records will be stored, accessed, and maintained to ensure compliance and facilitate audits.

Benefits of an Environmental Safety Plan:

Regulatory Compliance: Ensures that the organization meets all relevant environmental laws and regulations, reducing the risk of legal penalties.

Risk Management: Helps identify and mitigate environmental risks, protecting the organization, its employees, and the community.

Sustainability: Promotes sustainable practices that reduce environmental impact and contribute to long-term resource conservation.

Improved Public Image: Demonstrates the organization's commitment to environmental stewardship, enhancing its reputation among stakeholders and the public.

Cost Savings: Reduces costs associated with waste disposal, energy consumption, and environmental remediation through efficient resource management and pollution prevention.

5.3 Success Story

Here's a success story on environmental safety that showcases the impact of collective action and innovative solutions:

Success Story: The Revival of the Ganges River in India

Background:

The Ganges River, also known as the Ganga, is one of the most sacred rivers in India. It is not only a spiritual and cultural symbol but also a lifeline for millions of people who depend on it for drinking water, agriculture, and industry. However, over the decades, the Ganges became one of the most polluted rivers in the world due to industrial discharge, untreated sewage, plastic waste, and religious offerings.

The Problem:

By the early 2000s, the pollution levels in the Ganges had reached alarming levels. The river's water quality deteriorated to the point where it was unsafe for human use. The pollution threatened the health and livelihoods of millions, as well as the river's rich biodiversity, including the endangered Ganges river dolphin.

The situation became so severe that it prompted widespread concern, both within India and internationally. The challenge was enormous, given the river's vast length and the number of cities and industries along its banks.

The Response:

In 2014, the Indian government launched the **Namami Gange

Programme**, a comprehensive initiative aimed at cleaning and rejuvenating the Ganges. The program combined modern technology, traditional practices, and community participation to tackle the pollution problem. Key steps taken under this program included:

1. Sewage Treatment Plants (STPs):

- One of the major contributors to river pollution was untreated sewage from cities along the Ganges. The Namami Gange Programme prioritized the construction and modernization of sewage treatment plants. Over 150 STPs were built or upgraded to treat sewage before it could enter the river.

2. Industrial Pollution Control:

The program enforced stricter regulations on industries located along the Ganges. It mandated zero liquid discharge for certain industries, meaning they could not release any liquid waste into the river. Continuous monitoring systems were installed to ensure compliance.

3. Solid Waste Management:

To address the issue of solid waste, particularly plastic, the program involved extensive riverbank cleaning drives. Thousands of volunteers participated in these efforts, and awareness campaigns were launched to educate people about proper waste disposal.

4. Afforestation and Biodiversity Conservation:

The program also focused on afforestation along the riverbanks to prevent soil erosion and improve the river's ecosystem. Conservation efforts were initiated to protect the Ganges river dolphin and other species that depend on the river.

5. Public Awareness and Participation:

Recognizing the cultural and religious significance of the Ganges, the program engaged with religious leaders, NGOs, and local communities to promote environmentally friendly practices. This included the proper disposal of religious offerings and the use of eco-friendly materials.

6. Infrastructural Development:

In addition to pollution control, the program included the development of ghats (riverfront steps) and crematoria with modern facilities to reduce the environmental impact of traditional practices.

The Outcome:

The Namami Gange Programme has led to significant improvements in the water quality of the Ganges. By 2021, the dissolved oxygen levels in many parts of the river, a key indicator of water health, had improved substantially. Several stretches of the river that were previously classified as "severely polluted" saw remarkable recovery.

Community Involvement: The program's success is largely attributed to the active participation of local communities and the alignment of environmental goals with cultural practices. The widespread public awareness campaigns changed people's attitudes toward the river and inspired community-led initiatives to keep the Ganges clean.

Economic and Social Benefits: The improved health of the Ganges has had positive economic effects, particularly for agriculture and fishing communities. The rejuvenation of the river has also boosted tourism in several regions, leading to economic growth.

Biodiversity Restoration: Efforts to protect the river's biodiversity have shown encouraging results, with an increase in the population of the Ganges river dolphin and other species.

Global Recognition: The Namami Gange Programme has been recognized globally as a model for large-scale environmental restoration efforts. It has demonstrated how a combination of government action, community participation, and technological innovation can lead to significant environmental improvements.

Lessons Learned:

Integrated Approach: Environmental issues require a holistic approach that combines regulatory measures, technological solutions, and community involvement.

Sustained Effort: Long-term commitment and consistent effort are crucial for addressing environmental challenges. The success of the Namami Gange Programme is the result of continuous efforts over several years.

Cultural Integration: Aligning environmental initiatives with cultural values can lead to greater public support and sustainable outcomes.

This success story of the Ganges River's revival illustrates how determined action, supported by both government and communities, can lead to meaningful environmental safety outcomes, even in the face of seemingly insurmountable challenges.

5.4 Brainstorm

Here's a 31-question multiple-choice test on environmental safety:

Environmental Safety Multiple-Choice Test

1. What is the main goal of environmental safety practices?

   a) To reduce energy consumption

   b) To protect the environment and human health

   c) To increase industrial productivity

   d) To minimize waste production only

2. Which of the following is considered a hazardous waste?

   a) Plastic bottles

   b) Used batteries

   c) Organic compost

   d) Glass jars

3. What is the most effective method to manage waste?

   a) Burning

   b) Recycling

   c) Landfilling

   d) Dumping in the ocean

4. Which of the following gases is a significant contributor to global warming?

   a) Nitrogen

b) Oxygen

c) Carbon dioxide

d) Helium

5. What does the acronym "EPA" stand for in environmental safety?

   a) Environmental Pollution Agency

   b) Environmental Protection Agency

   c) Environmental Preservation Agency

   d) Environmental Prevention Agency

6. Which of these is NOT a renewable energy source?

   a) Solar power

   b) Wind power

   c) Coal

   d) Hydropower

7. What is the primary cause of deforestation?

   a) Urban development

   b) Agricultural expansion

   c) Climate change

   d) Mining

8. What is an example of a sustainable practice?

   a) Clear-cutting forests

   b) Overfishing

c) Using public transportation

d) Drilling for oil

9. Which of the following can help reduce air pollution?

   a) Using cars for short distances

   b) Burning leaves and trash

   c) Walking or biking instead of driving

   d) Using fossil fuels for energy

10. What is the term for harmful chemicals accumulating in the food chain?

    a) Bioaccumulation

    b) Biodegradation

    c) Biomagnification

    d) Bioengineering

11. Which of the following best describes "sustainable development"?

    a) Development without any environmental impact

    b) Development that meets present needs without compromising future generations

    c) Development that focuses solely on economic growth

    d) Development that uses all available resources

12. What does the term "carbon footprint" refer to?

    a) The amount of carbon in the atmosphere

    b) The total greenhouse gases produced by human activities

c) The area covered by forests

d) The impact of carbon emissions on marine life

13. Which of the following is an example of water conservation?

a) Letting the tap run while brushing your teeth

b) Fixing leaky faucets

c) Washing cars frequently

d) Watering lawns during peak sunlight hours

14. What is the most environmentally friendly way to dispose of organic waste?

a) Burning it

b) Composting

c) Throwing it in the trash

d) Flushing it down the toilet

15. Which of the following is a consequence of improper waste disposal?

a) Increased recycling

b) Soil contamination

c) Reduced energy consumption

d) Enhanced biodiversity

16. What is a common source of water pollution?

a) Solar energy

b) Industrial runoff

c) Wind turbines

d) Planting trees

17. What does "LEED" certification relate to?

    a) Energy-efficient appliances

    b) Sustainable building practices

    c) Renewable energy sources

    d) Eco-friendly transportation

18. Which of the following contributes to soil erosion?

    a) Planting cover crops

    b) Deforestation

    c) Sustainable farming

    d) Afforestation

19. What is a major source of indoor air pollution?

    a) Ventilation systems

    b) Plants

    c) Paints and cleaning products

    d) Natural light

20. Which practice can help reduce greenhouse gas emissions?

    a) Using single-use plastics

    b) Driving fuel-efficient vehicles

    c) Increasing energy consumption

    d) Deforestation

21. What is the purpose of an Environmental Impact Assessment (EIA)?

a) To assess the economic impact of a project

b) To evaluate the potential environmental effects of a proposed project

c) To measure pollution levels in a region

d) To certify eco-friendly products

22. Which of the following is a benefit of using renewable energy?

a) Depletion of natural resources

b) Reduction in greenhouse gas emissions

c) Increase in air pollution

d) Higher levels of water consumption

23. What does the "three R's" of environmental conservation stand for?

a) Reduce, Reuse, Recycle

b) Renew, Replant, Revive

c) Replenish, Restore, Recycle

d) Reduce, Recycle, Replenish

24. What is a common effect of oil spills on marine environments?

a) Enhanced fish populations

b) Improved water quality

c) Destruction of marine habitats

d) Increased tourism

25. Which of the following actions can help combat climate change?

    a) Increasing fossil fuel consumption

    b) Planting trees

    c) Reducing energy efficiency

    d) Expanding urban sprawl

26. Which chemical is primarily responsible for the depletion of the ozone layer?

    a) Carbon dioxide

    b) Methane

    c) Chlorofluorocarbons (CFCs)

    d) Nitrous oxide

27. What is an environmental benefit of using electric vehicles?

    a) Increased greenhouse gas emissions

    b) Reduced air pollution

    c) Depletion of natural resources

    d) Higher fuel consumption

28. Which of the following best describes "greenwashing"?

    a) Legitimate environmental practices

    b) Misleading claims about environmental benefits

    c) A process of cleaning polluted areas

d) A certification for eco-friendly products

29. What is a consequence of ocean acidification?

    a) Growth of coral reefs

    b) Decrease in marine biodiversity

    c) Increase in fish populations

    d) Improvement in water quality

30. What is one way to reduce plastic pollution?

    a) Using more plastic bags

    b) Burning plastic waste

    c) Reusing and recycling plastic products

    d) Dumping plastic in landfills

Answer:

1. b) To protect the environment and human health
2. b) Used batteries
3. b) Recycling
4. c) Carbon dioxide
5. b) Environmental Protection Agency
6. c) Coal
7. b) Agricultural expansion
8. c) Using public transportation
9. c) Walking or biking instead of driving

10. c) Biomagnification
11. b) Development that meets present needs without compromising future generations
12. b) The total greenhouse gases produced by human activities
13. b) Fixing leaky faucets
14. b) Composting
15. b) Soil contamination
16. b) Industrial runoff
17. b) Sustainable building practices
18. b) Deforestation
19. c) Paints and cleaning products
20. b) Driving fuel-efficient vehicles
21. b) To evaluate the potential environmental effects of a proposed project
22. b) Reduction in greenhouse gas emissions
23. a) Reduce, Reuse, Recycle
24. c) Destruction of marine habitats
25. b) Planting trees
26. c) Chlorofluorocarbons (CFCs)
27. b) Reduced air pollution
28. b) Misleading claims about environmental benefits
29. b) Decrease in marine biodiversity
30. c) Reusing and recycling plastic products

Part 6

Behaviour-based Safety

PART VI

6. BEHAVIOUR-BASED SAFETY

6.1 Introduction to BBS

Behaviour-Based Safety (BBS) is a proactive approach to safety management that focuses on identifying and influencing the behaviours of individuals to reduce the risk of accidents and injuries in the workplace. Unlike traditional safety approaches that emphasize compliance with safety procedures and the use of personal protective equipment (PPE), BBS targets the root cause of many safety incidents: human behaviour. The primary goal of BBS is to create a culture of safety where safe behaviours are encouraged and unsafe behaviours are addressed through observation, feedback, and positive reinforcement.

Introduction to Behaviour-Based Safety:

1. Concept and Principles:

Behaviour Focus: BBS is based on the understanding that many workplace accidents are caused by unsafe behaviours rather than equipment failures or environmental hazards alone. It emphasizes the need to observe and influence employees' actions to improve safety outcomes.

Behaviour Modification: By analysing and modifying behaviours, BBS aims to prevent accidents before they occur. This involves identifying risky behaviours, providing feedback, and reinforcing safe practices.

2. Core Components:

Observation: Employees and safety professionals observe behaviours in the workplace to identify both safe and unsafe actions. Observations are often structured and systematic, focusing on specific tasks and processes.

Feedback: After observing behaviours, feedback is provided to individuals. This feedback is constructive and aimed at reinforcing safe behaviours and correcting unsafe ones.

Reinforcement: Positive reinforcement is used to reward safe behaviours, creating motivation for employees to continue following safety practices. This can include recognition, rewards, or other incentives.

Employee Involvement: BBS encourages active participation from employees at all levels. Employees are involved in identifying risks, suggesting improvements, and taking ownership of safety practices.

3. Implementation Process:

Assessment: Begin by assessing the current safety culture and identifying areas where behaviour-based interventions can be applied. This may include analysing incident reports and safety data.

Training: Train employees and managers on BBS principles, observation techniques, and how to give effective feedback. Training ensures that everyone understands the importance of behaviour in safety.

Observation and Data Collection: Implement a system for observing and recording behaviours related to safety. This data helps identify trends and areas for improvement.

Feedback and Communication: Establish a process for providing feedback to employees and communicating the results of observations. Ensure that feedback is timely, specific, and supportive.

Continuous Improvement: Use the data collected to continuously improve safety practices and address any emerging issues. Regularly review and update the BBS program to maintain its effectiveness.

4. Benefits of Behaviour-Based Safety:

Reduced Accident Rates: By addressing unsafe behaviours and promoting safe practices, BBS helps reduce the frequency and severity of workplace accidents and injuries.

Enhanced Safety Culture: BBS fosters a culture of safety where employees are actively engaged in safety practices and take responsibility for their own and their colleagues' safety.

Increased Employee Involvement: Involving employees in safety initiatives empowers them to contribute to the safety program, leading to greater buy-in and compliance.

Improved Communication: The feedback and observation processes improve communication between employees and management, leading to better understanding and collaboration on safety issues.

5. Challenges and Considerations:

Resistance to Change: Some employees may resist BBS initiatives due to fear of criticism or a lack of understanding of the program's benefits. Overcoming this resistance requires clear communication and support from management.

Consistency and Fairness: Ensuring that observations and feedback are consistent and fair is crucial for maintaining trust and credibility in the BBS program.

Sustainability: Maintaining the effectiveness of a BBS program requires ongoing commitment and resources. Regular reviews and updates are necessary to keep the program relevant and effective.

6.2 Principals of Behaviour-Based Safety

Behaviour-Based Safety (BBS) is grounded in several key principles that guide its implementation and effectiveness. These principles focus on understanding and influencing human behaviour to improve safety in the workplace. Here are the core principles of BBS:

1. Focus on Behaviour:

Principle: Safe and unsafe behaviours are the primary causes of accidents and injuries in the workplace. Addressing these behaviours is essential for improving safety.

Application: Identify and observe specific behaviours related to safety rather than focusing solely on environmental or equipment-related issues.

2. Behaviour Observation:

Principle: Systematic observation of employees' behaviours provides valuable insights into safety practices and potential risks.

Application: Conduct regular and structured observations to identify both safe and unsafe behaviours, documenting them for analysis and feedback.

3. Feedback and Communication:

Principle: Providing timely and constructive feedback is crucial for reinforcing safe behaviours and addressing unsafe ones.

Application: Use observations to give feedback that is specific, non-punitive, and aimed at encouraging safe behaviour. Communicate effectively with employees to support their understanding and improvement.

4. Positive Reinforcement:

Principle: Reinforcing safe behaviours with positive rewards and recognition encourages employees to continue practising safety.

Application: Implement reward systems, recognition programs, or other incentives to acknowledge and reinforce safe behaviours, boosting morale and motivation.

5. Employee Involvement:

Principle: Involving employees in the safety process increases engagement and ownership of safety practices.

Application: Encourage employees to participate in identifying safety risks, suggesting improvements, and taking part in the observation and feedback processes.

6. Continuous Improvement:

Principle: BBS is a dynamic process that requires ongoing assessment and refinement to remain effective.

Application: Regularly review and update the BBS program based on feedback, incident data, and changing workplace conditions to ensure continuous safety improvement.

7. Safety Culture Development:

Principle: Developing a strong safety culture where safety is a shared value among all employees enhances the effectiveness of BBS.

Application: Foster a culture where safety is prioritized, open communication is encouraged, and employees feel responsible for their own and their colleagues' safety.

8. Management Commitment:

Principle: Successful BBS implementation requires strong support and commitment from management.

Application: Ensure that leadership actively supports and participates in BBS initiatives, provides resources, and demonstrates a commitment to safety.

9. Data-Driven Decisions:

Principle: Decisions and improvements should be based on data collected through observations and incident reports.

Application: Use data to identify trends, assess the effectiveness of interventions, and make informed decisions about safety practices.

10. Education and Training:

Principle: Educating and training employees on BBS principles, observation techniques, and safe behaviours is essential for program success.

Application: Provide comprehensive training for employees and managers on BBS concepts, effective observation, and feedback techniques.

11. Non-Punitive Approach:

Principle: BBS should focus on improving behaviours and safety, not on punishing individuals for unsafe actions.

Application: Address unsafe behaviours through constructive

feedback and support rather than punitive measures, fostering a positive environment for improvement.

Conclusion:

The principles of Behaviour-Based Safety emphasize the importance of focusing on behaviour, involving employees, and fostering a supportive safety culture to improve workplace safety. By adhering to these principles, organizations can effectively reduce accidents and injuries, enhance employee engagement, and create a safer work environment.

6.3 Behaviour-Based Safety (BBS) approach

Behaviour-Based Safety (BBS) is a proactive safety management approach that focuses on understanding and influencing employee behaviours to improve workplace safety. The core idea of BBS is that most workplace accidents and injuries result from unsafe behaviours rather than solely from environmental hazards or equipment failures. By targeting behaviours, BBS aims to prevent incidents before they occur, foster a culture of safety, and enhance overall safety performance.

Key Elements of the Behaviour-Based Safety (BBS) Approach:

1. Behaviour Focus:

Definition: BBS emphasizes the role of human behaviour in workplace safety. It recognizes that many safety incidents are caused by unsafe behaviours and that modifying these behaviours can reduce accidents.

Application: Identify critical safety behaviours that need to be reinforced and address those that are unsafe or risky.

2. Systematic Observation:

Definition: Systematic observation involves watching employees perform their tasks and recording their behaviours related to safety.

Application: Observations should be structured and regular, focusing on specific tasks and safety practices. This helps identify both safe and unsafe behaviours.

3. Feedback and Reinforcement:

Definition: Providing feedback is essential for influencing behaviour. Positive reinforcement encourages the continuation of safe behaviours, while corrective feedback addresses unsafe actions.

Application: Offer timely, constructive feedback and recognize safe behaviours with rewards or other forms of positive reinforcement.

4. Employee Involvement:

Definition: Engaging employees in the safety process increases their commitment to safe practices and helps build a safety culture.

Application: Involve employees in identifying safety risks, participating in observations, and suggesting improvements. Encourage ownership and responsibility for safety.

5. Data-Driven Approach:

Definition: BBS relies on data collected from observations and incident reports to inform safety practices and decision-making.

Application: Analyse observation data to identify trends, assess the effectiveness of interventions, and make informed decisions to improve safety.

6. Safety Culture Development:

Definition: Developing a strong safety culture where safety is a shared value helps integrate BBS principles into everyday work practices.

Application: Promote a culture where safety is prioritized, open communication is encouraged, and employees feel responsible for their own and others' safety.

7. Management Commitment:

Definition: Successful BBS implementation requires support and commitment from management.

Application: Ensure that leadership actively supports BBS initiatives, provides necessary resources, and participates in safety activities.

8. Continuous Improvement:

Definition: BBS is an ongoing process that requires regular review and adjustment to remain effective.

Application: Continuously assess the BBS program, update strategies based on feedback and data, and implement improvements to enhance safety performance.

9. Education and Training:

Definition: Training is crucial for the successful implementation of BBS. It ensures that employees and managers understand BBS principles and techniques.

Application: Provide training on observation methods, effective feedback, and safe behaviours to all employees involved in the BBS process.

10. Non-Punitive Approach:

Definition: BBS focuses on behaviour improvement rather than punishment. The goal is to support employees in adopting safe practices rather than penalizing them for unsafe actions.

Application: Address unsafe behaviours through positive

reinforcement and constructive feedback rather than disciplinary measures.

Benefits of the Behaviour-Based Safety Approach:

Reduced Accident Rates: By addressing unsafe behaviours, BBS helps decrease the number and severity of workplace accidents and injuries.

Enhanced Safety Culture: BBS fosters a safety-conscious culture where employees are actively involved in promoting and maintaining safety.

Increased Employee Engagement: Employees who participate in BBS initiatives are more likely to be engaged and committed to safety.

Improved Communication: The feedback and observation processes enhance communication between employees and management, leading to better safety practices.

Conclusion:

The Behaviour-Based Safety (BBS) approach is an effective strategy for improving workplace safety by focusing on the behaviours that lead to accidents and injuries. By systematically observing behaviours, providing feedback, involving employees, and fostering a safety culture, organizations can reduce risks, enhance safety performance, and create a safer work environment.

6.4 Success Story

A powerful success story about human life in terms of behaviour is the story of Malala Yousafzai, a young Pakistani girl whose courage and resilience transformed her from a victim of violence into a global advocate for education and women's rights.

The Background

Malala was born in the Swat Valley of Pakistan, an area that was eventually overtaken by the Taliban. Under their rule, girls were banned from attending school, which was something Malala, a passionate advocate for education, could not accept. She began speaking out publicly against this restriction, even writing a blog for the BBC under a pseudonym, sharing her experiences and advocating for girls' education.

The Incident

In 2012, when Malala was just 15 years old, the Taliban attempted to silence her by attacking her on her way home from school. She was shot in the head and critically injured. Most people would have given up after such a traumatic event, but Malala's behaviour demonstrated remarkable strength and resilience. Instead of retreating in fear, she used the incident as a platform to amplify her message.

The Recovery and Rise

Malala was flown to the United Kingdom for treatment, and after a long recovery, she continued her activism. Her behaviour exemplified determination and an unwavering commitment to her cause despite the dangers she faced. She continued to speak out on the importance of education for girls, and her story gained international attention.

The Impact

Malala's behaviour—her courage, persistence, and compassion—earned her numerous accolades, including the Nobel Peace Prize in 2014, making her the youngest-ever laureate at the age of 17. She founded the Malala Fund, an organization dedicated to ensuring girls around the world have access to 12 years of free, safe, and quality education.

The Lesson

Malala Yousafzai's story is a testament to the power of behaviour in shaping one's life and impacting the world. Her courage in the face of danger, her resilience in overcoming tremendous adversity, and her relentless pursuit of justice for others exemplify how one's behaviour can lead to profound success and change.

Her story inspires millions, demonstrating that even in the darkest of times, behaviour driven by strong principles and values can lead to remarkable outcomes, not just for oneself but for society as a whole.

6.5 Brainstorm

Here's a 30-question multiple-choice test on Behaviour-Based Safety:

Behaviour-Based Safety (BBS) Multiple-Choice Test

1. What is the primary goal of Behaviour-Based Safety (BBS)?

   a) To punish unsafe behaviours

   b) To identify and reinforce safe behaviours

   c) To reduce the number of safety meetings

   d) To eliminate the need for safety training

2. Which of the following is a key component of a BBS program?

   a) Financial incentives

   b) Employee observation and feedback

   c) Increased production quotas

   d) Strict enforcement of rules

3. In BBS, what is the purpose of observing employees?

   a) To find reasons to discipline them

   b) To understand and improve safety behaviours

   c) To reduce break times

   d) To monitor productivity

4. What is the first step in implementing a BBS program?

   a) Conducting safety audits

b) Defining safe and unsafe behaviours

c) Rewarding employees

d) Increasing work speed

5. Which of the following is NOT a characteristic of BBS?

   a) Focus on observable behaviour

   b) Encourages positive reinforcement

   c) Based on scientific principles

   d) Ignores organizational culture

6. Why is positive reinforcement important in BBS?

   a) It encourages employees to work faster

   b) It increases the likelihood of repeating safe behaviours

   c) It eliminates the need for safety equipment

   d) It reduces the need for supervision

7. What is an example of a leading indicator in BBS?

   a) Number of accidents

   b) Employee observations

   c) Incident reports

   d) Lost workdays

8. How often should safety observations be conducted in a BBS program?

a) Once a year

b) Only after an accident

c) Regularly and consistently

d) Never

9. What should be done when unsafe behaviour is observed?

   a) Ignore it

   b) Punish the employee

   c) Provide immediate feedback and coaching

   d) Increase the employee's workload

10. What is a common pitfall in BBS programs?

    a) Focusing on positive reinforcement

    b) Collecting data without taking action

    c) Involving all employees

    d) Regularly reviewing observation data

11. Who should be involved in a BBS program?

    a) Only managers

    b) Only safety officers

    c) All employees, from frontline workers to management

    d) Only the most experienced employees

12. What is a "behavioural observation checklist" used for in BBS?

    a) To track productivity

    b) To identify safe and unsafe behaviours during observations

    c) To calculate bonuses

d) To monitor break times

13. What role does management play in a BBS program?

    a) No role; it's employee-driven

    b) Supporting and modelling safe behaviours

    c) Only implementing disciplinary actions

    d) Reducing the budget for safety

14. What should be the focus of BBS training?

    a) Punishment of unsafe behaviours

    b) Understanding and changing behaviours

    c) Increasing work speed

    d) Reducing costs

15. What is an example of positive reinforcement in BBS?

    a) A safety bonus for accident-free months

    b) A warning letter for unsafe behaviour

    c) Increased workload for safe employees

    d) Ignoring safe behaviour

16. What is the purpose of a BBS observation report?

    a) To document and analyse observed behaviours

    b) To report on employee productivity

    c) To identify employees to be disciplined

    d) To reduce the number of safety meetings

17. How does BBS differ from traditional safety programs?

a) BBS focuses on behaviour rather than conditions

b) BBS ignores safety rules

c) BBS is only for management

d) BBS reduces the need for safety equipment

18. What is a key benefit of implementing a BBS program?

a) Increased employee engagement in safety

b) Reduced need for training

c) Increased production quotas

d) Elimination of all accidents

19. How can BBS contribute to a positive safety culture?

a) By ignoring safety issues

b) By encouraging open communication and feedback on safety

c) By reducing the number of safety meetings

d) By focusing only on the management's perspective

20. What is the most important factor for the success of a BBS program?

a) Employee participation

b) High fines for unsafe behaviour

c) Lengthy observation checklists

d) Minimal management involvement

21. What should be done with the data collected from BBS observations?

a) Stored without review

b) Analysed to identify trends and areas for improvement

c) Used only for disciplinary actions

d) Disregarded after the observation

22. How can BBS help reduce workplace accidents?

a) By focusing solely on productivity

b) By identifying and addressing unsafe behaviours before they lead to accidents

c) By increasing the workload of employees

d) By eliminating safety training

23. What is a common criticism of BBS programs?

a) They focus too much on behaviour and not enough on workplace conditions

b) They are too expensive to implement

c) They ignore management's role in safety

d) They are only effective in large organizations

24. What is the role of feedback in a BBS program?

a) To punish employees for mistakes

b) To provide constructive guidance and reinforcement for safe behaviours

c) To reduce the number of safety rules

d) To increase employee workload

25. How does BBS address at-risk behaviours?

a) By ignoring them

b) By identifying and modifying the behaviours through positive reinforcement and coaching

c) By immediately firing employees

d) By reducing the amount of safety equipment available

26. What is the relationship between BBS and safety culture?

a) BBS undermines safety culture

b) BBS is a tool to help build and strengthen safety culture

c) BBS has no impact on safety culture

d) BBS only benefits individual safety

27. Why is consistency important in BBS observations?

a) It ensures accurate data and reliable feedback

b) It increases the workload on observers

c) It reduces the number of observations needed

d) It allows for skipping unsafe behaviours

28. How should an organization respond to trends identified through BBS data?

a) Ignore the trends

b) Implement corrective actions and enhance training

c) Use the trends only for disciplinary actions

d) Delay any response to avoid disruption

29. What is a key factor in maintaining employee motivation in a BBS program?

    a) Regular punishment for unsafe behaviours

    b) Consistent and meaningful recognition of safe behaviours

    c) Reducing employee involvement in safety decisions

    d) Increasing the complexity of safety procedures

30. What is the ultimate measure of success for a BBS program?

    a) Decreased number of safety meetings

    b) Improved safety behaviour and reduced incidents

    c) Increased productivity

    d) Reduction in training costs

Answer:

1. b) To identify and reinforce safe behaviours
2. b) Employee observation and feedback
3. b) To understand and improve safety behaviours
4. b) Defining safe and unsafe behaviours
5. d) Ignores organizational culture
6. b) It increases the likelihood of repeating safe behaviours
7. b) Employee observations

8. c) Regularly and consistently
9. c) Provide immediate feedback and coaching
10. b) Collecting data without taking action
11. c) All employees, from frontline workers to management
12. b) To identify safe and unsafe behaviours during observations
13. b) Supporting and modelling safe behaviours
14. b) Understanding and changing behaviours
15. a) A safety bonus for accident-free months
16. a) To document and analyse observed behaviours
17. a) BBS focuses on behaviour rather than conditions
18. a) Increased employee engagement in safety
19. b) By encouraging open communication and feedback on safety
20. a) Employee participation
21. b) Analysed to identify trends and areas for improvement
22. b) By identifying and addressing unsafe behaviours before they lead to accidents
23. a) They focus too much on behaviour and not enough on workplace conditions
24. b) To provide constructive guidance and reinforcement for safe behaviours
25. b) By identifying and modifying the behaviours through positive reinforcement and coaching

26. b) BBS is a tool to help build and strengthen safety culture
27. a) It ensures accurate data and reliable feedback
28. b) Implement corrective actions and enhance training
29. b) Consistent and meaningful recognition of safe behaviours
30. b) Improved safety behaviour and reduced incidents

TAKE AWAY – Last but foremost

To be remember the following things and insist to apply every day in your life.

Safety first then all.

Don't be a fool. Use the proper tool.

Turn your attention to accident prevention.

Don't be Safety blinded, be Safety minded.

Safety is Gainful, Accident is Painful.

Passion for Safety then no feel Courtesy.

Don't learn safety by accident.

Short cut leads to reduce your life.

There is no age bar to adopt Safety.

Work Safely Rule Proudly

Safety is our identity and we respect it

Our Goal—Zero Harm

==

New Beginning

==

Made in the USA
Coppell, TX
05 November 2026

11307220R00125